TENSIONS IN THE TERRITORIAL POLITICS OF WESTERN EUROPE

Tensions in the Territorial Politics of Western Europe

Edited by
R.A.W. RHODES and VINCENT WRIGHT

FRANK CASS

First published in 1987 in Great Britain by
FRANK CASS AND COMPANY LIMITED
Gainsborough House, Gainsborough Road,
London E11 1RS

and in the United States of America by
FRANK CASS AND COMPANY LIMITED
c/o Biblio Distribution Center
81 Adams Drive, P.O. Box 327, Totowa, NJ 07511

Copyright © 1987 by Frank Cass & Co. Ltd.

British Library Cataloguing in Publication Data

Tensions in the territorial politics of
Western Europe.
1. Territory, National—Europe
2. Regionalism—Europe 3. Europe—
Politics and government—1945
I. Rhodes, R.A.W. II. Wright, Vincent
III. West European politics
320.1'2'094   JX4085

ISBN 0-7146-3329-1

This group of studies first appeared in a Special Issue on 'Tensions in the Territorial Politics of Western Europe' of *West European Politics*, Vol. 10, No. 4, October 1987, published by Frank Cass & Co. Ltd.

All rights reserved. No part of this publication may be reproduced, stored in a retrieval system or transmitted in any form or by any means, electronic, mechanical, photocopying, recording or otherwise, without the prior permission of Frank Cass and Company Limited.

Printed and bound in Great Britain by
Adlard & Son Ltd, The Garden City Press

Contents

Notes on the Contributors		vi
Editors' Preface		viii
Introduction	*R. A. W. Rhodes and Vincent Wright*	1
Territorial Politics in the United Kingdom: The Politics of Change, Conflict and Contradiction	*R. A. W. Rhodes*	21
France: The Construction and Reconstruction of the Centre, 1945–86	*Yves Mény*	52
The Federal Republic of Germany: From Co-operative Federalism to Joint Policy-making	*Joachim Jens Hesse*	70
Italy – Territorial Politics in the Post-War Years: The Case of Regional Reform	*Robert Leonardi, Raffaella Y. Nanetti and Robert D. Putnam*	88
The Netherlands: A Decentralised Unitary State in a Welfare Society	*Theo A. J. Toonen*	108
Ireland: The Interplay of Territory and Function	*T. J. Barrington*	130
The West European State: The Territorial Dimension	*L. J. Sharpe*	148
Abstracts		168

Notes on Contributors

R. A. W. Rhodes is Reader in Government, University of Essex; Visiting Professor, European Institute of Public Administration, Maastricht, The Netherlands; and joint editor, *Public Administration* (journal of the Royal Institute of Public Administration). His books include (as co-author) *Intergovernmental Relations in the European Community* (1977); *Public Administration and Policy Analysis* (1979); *Control and Power in Central Local Government Relations* (1981); (as co-author) *The New British Political System* (1983); and *The National World of Local Government* (1986). He has published articles on local government in Britain and Europe, regional government, the development of public administration and the sociology of organisations. His current research is on bureaucratic politics in Britain and territorial politics in Western Europe.

Vincent Wright is a Fellow of Nuffield College, Oxford. He is the author of *The Government and Politics of France* (1978) and of books on the French Prefects and the Council of State. With Gordon Smith, he edits *West European Politics*, and he has also edited or co-edited several works on various aspects of French, and of European, politics.

Yves Mény is a Professor at the University of Paris II and Director of the Groupe de Recherches coordonnées sur l'administration locale. He was Professor at the European University Institute in Florence. His publications include: *Centralisation et Décentralisation dans le débat politique français* (1974), *Dix ans de Régionalisation en Europe* (1982), *Centre-Periphery Relations in Western Europe* (with Vincent Wright, 1985), *The Politics of Steel: Western Europe and the Steel Industry in the Crisis Years* (with Vincent Wright, 1986).

Theo A. J. Toonen is Associate Professor at the Erasmus University, Rotterdam, Department of Public Administration. His teaching and research interests include topics on intergovernmental relations, the organisation of metropolitan areas, and theoretical developments in public administration. His recent publications cover topics such as organisational political approaches to decentralisation, central–local relationships in the Netherlands, intermunicipal co-operation as a form of territorial reorganisation, transfer of knowledge and information from national to local authorities and historical developments in the political theory of the Dutch decentralised unitary state.

L. J. Sharpe is a University Lecturer in Public Administration and Fellow of Nuffield College, Oxford, since 1965. From 1966 to 1969 he was Research Director for the Royal Commission on Local Government for England and he edited *Political Studies* from 1976 to 1982. His publications include *A Metropolis Votes* (1963), and *Does Politics Matter?* (1984). He has also edited two volumes: *Decentralist Trends in Western Democracies* (1980) and *The Local Fiscal Crisis: Myths and Realities* (1981).

Joachim Jens Hesse is Professor of Political and Administrative Sciences at the Post-Graduate School of Administrative Sciences, Speyer, Federal Republic of Germany. He has published extensively in the areas of state and administrative sciences, federalism, urban affairs and regional studies. Recent publications include: *Erneuerung der Politik 'von unten'? Stadtpolitik und kommunale Selbstverwaltung im Umbruch* (1986), and (with Thomas Ellwein, Renate Mayntz and Fritz W. Scharpf), *Jahrbuch zur Staats- und Verwaltungswissenschaft* (1987).

Robert Leonardi, Associate Professor of Political Science at DePaul University and Jean Mommet Fellow at the European University Institute, is the co-editor of *Italian Politics, A Review* and has published work on Italian political institutions at the national and sub-national level since 1972.

Raffaella Y. Nanetti, Associate Professor of Urban Planning and Policy at the University of Illinois at Chicago, is co-editor of *Italian Politics, A Review* and has done extensive work on urban and economic planning in Italy and other advanced, industrialised countries.

Robert D. Putnam, Professor of Government at Harvard University, has written extensively on Italian and international affairs. His latest book is *Hanging Together: the Seven Powers Summits* (1987).

T.J. Barrington was a senior civil servant in the Department of Local Government (now Environment), Dublin, from 1944 to 1960. He was a founder and first Director, Institute of Public Administration, Dublin, 1960–76. He was Editor of *Administration*, 1953–63, and a member of the Devlin Inquiry into the Irish Civil Service, 1966–69. Author of *From Big Government to Local Government*, Dublin (1975) and *The Irish Administrative System* (1980), he has written and lectured extensively on government and administration. He has been a member of the Executive Committee, International Institute of Administrative Sciences, 1971–80, and of the European Group of Public Administration, 1973–80.

Editors' Preface

Any collective endeavour incurs many debts. This collection's origins lie in a seminar on 'The Construction and Reconstruction of Centers and Center–Periphery Relations in Europe' held at the Israel Academy of Sciences and Humanities (Jerusalem) in June 1984. A number of the participants at that seminar shared an interest in territorial politics, and took part in a second, smaller workshop on 'Territorial Politics in Western Europe' which was held at Nuffield College (Oxford) in October 1985. Both meetings were held under the auspices of the European Science Foundation (ESF) and funded for the UK participants by the Economic and Social Research Council (ESRC). We should like to express our gratitude to both institutions for their support. Walter Rügg and his colleague Elisabeth Luterbacher (Bern) helped to organise the Jerusalem seminar and provided encouragement and help for the Oxford workshop. All participants are grateful for their endeavours, none more so than the editors who hope this collection is some small recompense for their hard work.

A number of individuals contributed to the Oxford workshop. We would like to thank Jim Bulpitt (Warwick), Sabino Cassesse (Rome), the late Sir Norman Chester (Nuffield College), Pat Commins (Dublin), Hans Daalder (Leiden) and Nevil Johnson (Nuffield College) for their interest, criticism and advice. Rod Rhodes would like to thank Carol Snape for her accurate and prompt typing of the manuscript. Vincent Wright would like to thank the Warden and Fellows of Nuffield College who hosted the second conference, and Trude Hickey whose secretarial assistance proved so invaluable. Finally, both editors would like to thank the authors for promptly revising manuscripts and responding to detailed queries.

R.A.W. Rhodes, University of Essex
Vincent Wright, Nuffield College, Oxford

Introduction

R. A. W. Rhodes and Vincent Wright

To study 'territorial politics' is to raise the problems of what is being studied and how it is to be understood. This introduction aims to avoid obscurity of intent and conception. At the outset, therefore, the subject matter is defined and the objectives and themes of this collection are outlined. Thereafter, and with more regard for nuance and ambiguity, the problems raised by our collective endeavours for possible future work in the field are discussed. The advantages and difficulties of incorporating comparative and historical dimensions into the study of intergovernmental relations (IGR) − from which amalgamation comes the appellation 'territorial politics' − are explored. The studies of individual countries provide ample illustration of both the advantages and difficulties. It would be presumptuous, of course, to claim that the differences between the several systems of territorial politics in Western Europe are explained in this collection. None the less, it identifies a range of factors which need to be included in any such 'grand design' and suggests future topics for research which may aid any 'master-builder'.

OBJECTIVES AND THEMES

The 'what' of this volume is perhaps most commonly referred to as 'centre−periphery relations', but this concept is magnificently vague. Bulpitt[1] provides a useful conspectus of approaches: he distinguishes between territorial systems analysis, the centralisation/decentralisation dichotomy, the internal colonialism thesis, and centre−periphery relations proper. This latter approach is further sub-divided into diffusion theory, territorial dependency and administrative/political brokerage.[2]

Territorial systems analysis focuses on the legal-institutional classification of political systems: e.g. the distinction between confederal, federal and unitary systems.[3] The centralisation thesis focuses on measuring the concentration of decision-making in central government and often criticises the adverse consequences of such concentration, especially the decline of local autonomy.[4] The internal colonialism thesis argues that economically-advanced areas colonise − that is, dominate and exploit − less advanced areas: peripheral regions are dependent on the centre.[5] This approach is also seen as a major strand within the fourth (and final) approach to territorial politics, namely centre−periphery relations. Additionally, Bulpitt discusses diffusion theory which views the distinction between centre and periphery as a fundamental (and major) social cleavage eroded in the process of modernisation or nation-building,[6] and brokerage theories which focus on the interactions between the administrative and political élites of centre and periphery.[7]

If this plethora of approaches is deemed insufficiently daunting (and

Bulpitt's scheme is illustrative, not exhaustive), a brief note of the range of 'problems', or substantive research topics, associated with them should intimidate even the most enthusiastic reader. Even in the literature on Britain it includes the persistence of nationalism and the rise and fall of nationalist parties; the demise of local government; the modernisation of territorial institutions; intergovernmental relations; and the evolution of the British 'Union'. Subsuming this variety of preoccupations under the single label of 'centre–periphery relations' serves only to obscure the inordinately diffuse character of the field. A narrower theoretical focus and a more specific subject matter is required if the field is not to become all things to all men. This collection aims to provide a greater degree of precision through a focus on territorial politics.

Bulpitt defines territorial politics broadly as: 'that arena of political activity concerned with the relations between central political institutions in the capital city and those interests, communities, political organisations and governmental bodies outside the central institutional complex but within the accepted boundaries of the state, which possess, or are commonly perceived to possess, a significant geographical or local/regional character'.[8] This label has at least one advantage: it avoids a concentration on relations between central and local government. All sub-central units of government are included. There are, however, disadvantages. As defined by Bulpitt the concept is so broad that it becomes difficult 'to separate central-local relations from the wider workings of the political system', and 'Everything will appear to be connected ... the distinctiveness of the territorial political arena will be obscured'.[9] This problem arises from the inclusion of any 'interest, communities ... commonly perceived to have a significant geographical or local/regional character' in the definition. Nor is this the only problem in Bulpitt's formulation.[10] A narrower definition of territorial politics avoids some of these problems. Here it refers to *'the arena of political activity concerned with the relations between central political institutions in the capital city and those sub-central political organisations and governmental bodies within the accepted boundaries of the state'*. In this way, it is possible to retain a broad institutional compass and still distinguish the topic from the wider political system. However, it might be argued that this definition also makes the study of territorial politics indistinguishable from the study of intergovernmental relations.

Of all the approaches encompassed by the term centre–periphery relations, intergovernmental theory is seen as the most restricted because it is said to focus on institutional relationships. Setting this criticism to one side for the moment, the key difference between the study of territorial politics and of intergovernmental relations lies not in the latter's concern with institutions but in the former's recognition that there is a need 'to return to history'.[11] A focus upon territorial politics fuses, therefore, the concerns of intergovernmental theory with history.

The reasons for some focus on history are threefold. In the first place, contemporary events have to be located historically if cause and effect continuities and discontinuities are to be accurately distinguished. Secondly, interpretations of the past are part of the stuff of contemporary political debates and such

INTRODUCTION

interpretations need to be tested or contested. Thirdly, many theoretical issues can be explored only through historical analysis: 'history is too important to be left to historians'.[12]

There are two possible ways in which historical analysis can be incorporated into the study of territorial politics: synoptic, synthetic surveys and the analysis of critical junctures. The former spans the decades, even centuries, identifying general trends and providing an interpretive backcloth for current events. The latter focuses on either particular events or years deemed to be turning points in the development of territorial politics, documenting the events of concern and demonstrating their continuing impact.

Ideally, of course, the synoptic surveys should encompass the gestation of the modern state in the nineteenth century, but this is no easy task, and is beyond the scope of this work.[13] However, even a more modest historical prospectus has marked advantages. Too often, the study of territorial politics is an exercise in 'upper class journalism' or, to be less tendentious, is over-preoccupied with current affairs. Covering the post-war period represents, therefore, a useful step, if only a first step, in broadening the perspective. Attention is directed away from personalities and burning issues of the moment towards longer trends and the role of the structural characteristics of the system of territorial politics in shaping current issues and responses. Political scientists may have ignored or forgotten, as Jim Sharpe suggests in his contribution, the territorial aspects of politics. Even the relatively restricted focus of the post-war years is a timely reminder that territorial questions, in a multiplicity of forms, lie at the heart of the emergence and development of the modern nation state. A recurrent theme in all the following contributions is that functional politics in the guise of the welfare state has not eclipsed territorial politics, merely induced new forms of interaction (and conflict) between the two.

To return to the relationship between territorial politics and intergovernmental relations, it is crucially important to note that the latter term is not limited to institutional relationships (and most definitely not to the study of central–local relations). It encompasses all 'sub-central political organisations and governmental institutions' and it concentrates on the processes of administrative and political brokerage. Moreover, it is precisely these relationships and processes which are of key importance in the territorial politics of advanced industrial societies.

Western Europe has experienced a massive increase in the scale of government intervention in the post-war period, and one central element in such growth has been the development of the welfare state. Moreover, throughout Western Europe, sub-central units of government have often been the prime vehicle for providing these services. Governments have faced the dilemma, therefore, of being electorally accountable for services delivered through differentiated, disaggregated policy systems. The territorial allocation (or distribution) of functions has become a major political and policy issue and, under the pressures of world recession, all centres have developed strategies for regulating the function-based policy system. Any exploration of this topic involves the analysis of policy networks, policy slippage and policy inertia (for a somewhat longer summary see the contribution by Rhodes below).

Policy Network

The term 'central (or federal) government' is misleading because it implies a homogeneity of interest and purpose rarely present in reality. The term 'policy network' is a more appropriate metaphor for the analysis of national policy-making because it suggests that there are multiple function-specific networks at the centre, and it focuses attention on the relationships between organisations within the networks. Analysis is directed towards a comparison of networks — to the degree of integration within, and the degree of articulation between them.

Policy Slippage

This concept refers to the disparity between intentions and achievements. Thus central policy goals, as enacted, are often not realised because sub-central units of government modify the policy substantially in the process of implementation.

Policy Inertia

Confronted with 'resource squeeze'[14] governments have found that their ability to regulate disaggregated policy systems is severely constrained. Those interests directly benefiting from government policy, most notably professional groups, have become institutionalised in the structure of government and are, therefore, able to resist attempts to curtail the rate of growth in expenditure and cuts in public expenditure. The several policy systems are self-regulating and are able to resist the unilateral and authoritarian exercise of central authority — even in systems which are categorised as highly centralised.

Those particular topics are not particularly novel and reflect a bias towards the management of intergovernmental relations. However, the effects of government intervention are pervasive. The growth of the welfare state has institutionalised professional interests and created client groups for the services. For instance, the provision of public housing has created a client group for this government service with a considerable interest in rent levels and the provisions for selling the property to tenants. This consumption (of public goods) cleavage may cross-cut other more traditional cleavages such as class. In short, state intervention in the form of the welfare state may create new social cleavages. Furthermore, central actors concerned to defend particular services can marshal the client group to protest about 'cuts' and to raise the spectre of electorally damaging policy implications — a tactic described as 'shroud-waving' when deployed by doctors in defence of health services. To understand centre–periphery relations in advanced industrial societies it is necessary to explore not only the changing forms of government intervention — i.e. strategies of regulation — but also the cleavages engendered by that intervention. Function *not* territory has become the basis of social cleavages.

Although cleavages between centre and periphery may persist (as the Spanish and Belgian cases clearly underline) they are being increasingly supplanted by functional divisions (which may, admittedly, be manifested territorially). This does not mean that the territorial dimension of politics is

INTRODUCTION

unimportant: as Jim Sharpe argues in his contribution, territoriality must remain a central preoccupation of political science. Rather it is to recognise that one of the key features of centre–periphery cleavages is that they are often divisions of, and in, the centre: professionalised central élites competing for scarce resources. Conflict and competition are no longer between centre and periphery but *within* the centre, between vertical coalitions of function-specific interests (central and sub-central). Accordingly, the description 'centre–periphery relations' is not only magnificently vague but it may also focus exclusively on cleavages which were significant in the evolution of the modern welfare state but which may be of declining relevance for the analysis of advanced industrial societies.

Intergovernmental theory as defined here is no longer concerned primarily with institutional analysis. It points to a set of rather different topics or questions:

- What are the variations in the territorial allocation of functions (i.e. between central/sub-central, elected/ad hoc governments)?
- What strategies are deployed by the centre to regulate function-specific, disaggregated policy systems?
- Have professional interests become institutionalised in the structure of government? To what degree does professional interest vary across specific policy issues and within, and between, policy networks?
- Has government intervention generated consumption cleavages? What is the relationship between professionalised central élites and their client groups?
- What are the distributional consequences for citizens, in terms of both optimal service delivery and democratic accountability, of variations between policy networks and in professional influence?

To be more precise, the proper focus of research ought to be functional allocation, interest articulation, central regulation and the distributional consequences of policy networks in the advanced industrial societies.

The following contributions explore these topics and demonstrate that this redefinition of intergovernmental theory may facilitate an understanding of the policy process in advanced industrial societies. Moreover, allied to historical analysis, it shows how these processes are a product of both the shared imperatives of policy-making in disaggregated systems and the distinct historical context of each country in the development of its governmental structure and welfare services. If centre–periphery relations have been concerned primarily with the development of the modern nation state, the approach in this work embraces a developmental approach to functional allocation in, and the policy processes of, the welfare state. The label 'territorial politics' is thus used to describe the amalgamation of history and intergovernmental theory.

In sum, the objectives of this work are:

- to argue for a focus on 'territorial politics';
- to demonstrate that the analysis of territorial politics requires an exploration of the historical context;

– to analyse major contemporary problems in territorial politics, focusing on functional allocation, interest articulation, central regulation and the distributional consequences of policy networks.

To achieve these objectives, the collection compares the territorial politics of two major unitary states – France and the United Kingdom; two major decentralised states – the Federal Republic of Germany and 'Regionalised' Italy; and two small European democracies – Ireland and the Netherlands. There are unfortunately, important omissions; there are no chapters on either Scandinavia or the 'new democracies' of Spain, Greece and Portugal. And the ever topical case of Belgium is also absent. None the less, this collection does encompass a range of territorial systems, providing suitably strenuous terrain for a comparative study in a field which, as yet, has attracted few explorers.[15]

PROBLEMS OF COMPARISON

To claim that all the authors in this collection worked within an agreed framework would be to offer a hostage to fortune. No collective endeavour worth its salt is characterised by anything but criticism, debate and agreement to differ. It is all the more surprising to record, therefore, that there was a substantial amount of common ground. Inevitably, it took a negative form: what not to do. Analysis had not to be confined to the legal-institutional characteristics of the system; to elected local authorities; and to comparing/ measuring the (elusive) degree of centralisation/decentralisation. More positively, it was agreed that the country studies had to encompass linkages and transactions between the centre and the *range* of sub-central government agencies and must explore the impact of sectoralisation (of functional politics) on such relationships. In effect, the contributors were prepared to work with the definition of territorial politics given above, reserving, of course, their right to criticise and amend the definition in the light of the problems uncovered in their specific country study. Consequently, the intellectual lineage of this work descends directly from the bureaucratic-bargaining model of centre– periphery relations.

To identify a hard core to our deliberations does not mean that they were confined to the core. However, before considering the limitations of our approach, it would be as well to demonstrate that it has some utility. It is possible to make a number of generalisations about the developments in, and the problems of IGR and territorial politics in Western Europe.

The problems of IGR are, to a remarkable degree, a reflection and a result of the problems of national economic management. With the onset of inflation in the 1970s, the 'fiscal crisis' was 'discovered', and governments began to worry about the issues of local expenditure and IGR. As Sharpe notes, this 'fiscal crisis' was compounded of rising local expenditure and an inelastic tax base but, above all, it was a problem of inflation.[16] Then, as now, the tensions of European IGR hinge on the national economic context and a government's diagnosis of the country's economic ills and not on the intrinsic features of the system of IGR.

Moreover, the definition of economic ills emphasises the growing scale of welfare expenditures. With inflation and world recession, these expenditures are seen as part of the general problem, since they pre-empt resources which would otherwise be available for the private sector to fuel economic recovery: hence the interest in curtailing public expenditure and in privatisation. Sub-central units, especially local authorities, have commonly been the preferred means for the expansion of welfare services. They are thus vulnerable to this reassessment of economic policy when the perceived needs of the private sector become paramount. Indeed, local authorities are doubly vulnerable. The expansion of services was substantially influenced by central governments. New services were not locally determined and, as a functional tool of the centre, local authorities found that service expansion went hand in hand with increased central supervision and regulation. Therefore, when the economic climate changed, the means for control already existed. Functional responsibility and local discretion were not commensurate, and the imbalance renders local government vulnerable to central intervention.

Nevertheless, the strategy of off-loading expenditure cuts to the periphery encountered the problems of the interdependence of centre and locality as well as the disaggregation of policy systems which are features of all the systems of IGR analysed in this work. Governments have been caught between the Scylla of policy slippage and the Charybdis of recalcitrance: if they bargain for compliance then policy is adapted and modified on implementation, at times substantially; if direction is preferred they have to confront local authority recalcitrance and the multiplication of unintended consequences.

In Western Europe, reactions to the problems of IGR were not uniform but were shaped by each country's political traditions. The form of, and changes in, each country's national government environment must constitute the heart of relevant historical debate, and the analysis of policy networks requires an exploration of their social, economic and political context. Indeed, there are significant points of contrast at the level of policy networks. Thus, local government systems differ in their range and type of responsibilities (or functions) and such variations obviously have a marked influence on the number and complexity of the policy networks in which local authorities are embedded. Even within the UK, there are sharp differences between local authorities in Northern Ireland and in England, with the former having a limited range of responsibilities. There is also the well-known difference between the tiny rural authorities and the major urban local authorities in France. The truly safe conclusion which emerges from the country studies is that there is considerable variation in the distribution of functions. Local authorities may have been, and remain, the leading edge of the welfare state, but there are still significant variations in the range of specific functions and their distribution between the types of sub-central authorities, not to mention the degree of sub-central discretion. Such comparisons may not tell us a great deal, however, and it is more revealing to examine the characteristics of policy networks.

The importance of local government's policy networks will vary with the role of central government. In Britain, the Netherlands and the Federal Republic of Germany (FRG) the non-executant role of the centre increases, other things being equal, its dependence on local authorities. The networks become, therefore, a crucial vehicle for central influence on policy implementation: indeed, the non-executant role provides considerable impetus for network building by the centre. To the extent that the centre has its own field agents, such 'hands-on' control decreases dependence and, indeed reverses roles because, as in the case of France, the centre can sometimes deliver services such as public works which are desired by the localities.

And, in spite of the exclusive focus on local government to this point, neither they nor field agencies are the only means of service delivery. The tools of government are multifarious, and the terminology for describing it is equally varied. Such 'fringe bodies' – to use the most opaque and, therefore, most general phrase – have been widely deployed and add both to the complexity of networks and to the problems of control. The prominence of the 'private initiative' in Dutch welfare society (not *state*) and the extraordinary range of parastatal bodies in France are perhaps the clearest illustrations of the point that analysis must encompass not only the range of sub-central authorities but also forms of collaboration with the private sector. The membership and interests of networks are complex, control varies in its efficiency and the problem of sub-optimisation or 'policy slippage' is universal, if not acute, in all systems of IGR.

The foregoing points emphasise variations in the fragmentation of governmental structure and the disaggregation of policy systems. But a comparison of patterns of disaggregation is only one aspect of the problem. Another aspect is a comparison of patterns of integration, and key agents in such integration are the professionals. The account of IGR in Britain stresses the importance of the institutionalisation of professions in the structure of government and this phenomenon recurs throughout Western Europe. Nor is integration a product solely of institutionalisation: it can be generated by shared values and ideology.[17] Such notions as 'best professional practice' can have a pervasive, if diffuse, effect. Moreover, irrespective of size, differentiation or specialisation is ubiquitous. Ireland, which is the smallest country of those studied here, appears to be the most centralised in Western Europe and yet its policy-making is highly fragmented and specialised. The Irish case illustrates the dangers of assessing territorial systems by comparing degrees of centralisation. Such an assessment, with its presumption that there is a unitary central actor, obscures a key facet of many advanced industrial societies; *the co-existence of fragmentation and centralisation*. Divergent interests within a centre, coupled with the professionalisation of functional policy systems, create multiple centres and erode horizontal co-ordination. In Luhmann's provocative phrase we live in an era of 'centreless' societies.[18] Each policy system may be centralised, however, at least in the sense of its centre repeatedly intervening in, if not controlling, the affairs of sub-central authorities. There is a persistent tension between the centre's desire to control and its capacity to do so – between its authority and its dependence: divergent interests *within* the centre are as important a constraint as the influence of interests *outside* it, because they may be carefully exploited by knowledgeable and politically-skilled localities.

Integration can also be generated by elected political actors. As in Britain, especially before the mid-1970s, a central government may actively support an intermediate tier of representation based on 'the national community of local government'. Local political élites were marginalised, as access to government decision-making was presaged on the aggregation of local government interests. Of the other countries in this work only the Netherlands has an equivalent sharp separation of national and local élites. Indeed, in Ireland, the brokerage role of parliamentarians fills the gap left by the emasculation of local political élites. However, the sharpest contrast is between Britain and France and the FRG.

The latter two countries are characterised by a honeycomb of political relations, with France the archetypal case. Its system of *cumul des mandats*, even though attenuated somewhat by the Socialist government's legislation, exemplifies the interpenetration of national and local élites, administrative and elected actors. Political channels of articulation are of pre-eminent importance and arguably foster substantial inertia, as the failure of the several attempts radically to reform local government before 1981 attests.[19]

At the organisational level, the resources of local governments vary, and this influences their ability to bargain with the centre. Such 'games' provide the 'local colour' in IGR and may be an important mechanism for enhancing the discretion of local government. But, and the proviso is crucial, an *exclusive* focus on these interactions inhibits comparative study because it takes as given the very factors which need to be explained in comparative analysis: namely, to cite but one example, the 'rules of the game'. Accordingly, while the analysis of bargaining games is an essential part of the analysis of IGR, the games must be contextualised: i.e. located within policy networks and their national environment.

In spite of the cursory nature of this discussion, intergovernmental theory does identify a range of problems common to IGR in West European liberal democracies. And it is important to note the distinctive nature of these problems. Little has been said about institutional reorganisation, central grant or financial controls. These facets of IGR are important, and are also widely discussed in the IGR literature. Other facets of IGR are equally important. This discussion has stressed interdependence, disaggregation, professionalisation and linkages (or systems of interest articulation) because these features of IGR are crucial to explaining resistance to change and inertia. Systems of IGR can be peculiarly resistant to central interventions. The causes and consequences of 'recalcitrance' – a term with markedly centralist overtones – is certainly a major problem to be grappled with.

To demonstrate that intergovernmental theory has its uses in the comparative study of territorial politics is not to turn a blind eye to its defects. Four problems require elucidation: the boundaries of government; complexity; accountability; and interests.

By extending the subject matter from the study of the relationship between central ministries and local authorities to the centre's relationship with *all* species of sub-central governmental and political organisations, the boundary between government and society becomes blurred almost to the point of becoming non-existent. From the disciplinary standpoint of political science, this extension raises a number of difficulties. For instance, the label 'government' seems inappropriate for several species of sub-central agency, especially those bodies with substantial private sector involvement. The activities and relationships under investigation do not seem to be distinctively governmental. Whether or not the subject matter should be seen as the domain of political science, it is clear that an exclusive focus on governmental institutions is profoundly misleading. If politics is about 'who gets what, when, how', territorial politics is concerned with 'who gets what public services, when, where, how and why'. It necessarily covers all those agencies which provide services on behalf of government as well as those government agencies directly providing services. This means that answering questions about the territorial distribution of public services is a multi-disciplinary endeavour. For citizens, access to, and the quality of, a service are of far greater importance than its parentage. The student of territorial politics must similarly look beyond the walls of disciplinary domesticity.

It is a truism to describe advanced industrial society as complex, yet if this description amounted to no more than claiming that government is complicated, it would still have some substance. However, as Elgin and Bushnell argue, to focus on the growth in complexity is already to identify a range of salient problems for advanced industrial society.[20] These problems include:

- Diminishing relative capacity of the individual to comprehend the overall system;
- Diminishing level of public participation in decision-making;
- Declining public access to decision-makers;
- Growing participation of experts in decision-making;
- Disproportionate growth in costs of co-ordination and control;
- Increasingly de-humanised interactions between people and the system;
- Increasing levels of alienation;
- Growing challenges to basic value premises;
- Increasing levels of unexpected and counterintuitive consequences of policy action;
- Increasing system rigidity;
- Increasing number of uncertainty of disturbing events;
- Narrow span of diversity of innovation;
- Declining legitimacy of leadership;
- Increasing system vulnerability;
- Declining overall performance of the system;
- Growing deterioration of the overall system unlikely to be perceived by most participants in that system.

There is no need to agree with each of these propositions in order to recognise that the appellation 'complex' is neither trite nor a truism; indeed, the more appropriate criticism would be that the word has to carry too heavy a burden of meaning.

The inadequacies of intergovernmental theory for the analysis of political accountability in disaggregated policy systems require a more extended discussion. Dunleavy has argued that this species of neo-pluralism makes over-optimistic assumptions about the consequences of differentiation.[21] Thus, differentiation promotes expertise in policy-making; fragments administrative jurisdictions, thereby requiring bargaining and negotiation between agencies; deconcentrates governmental functions, thereby reducing the degree of centralisation; and supplements accountability to elected assemblies with control by professional peer groups and their ethic of 'social responsibility'. Such outcomes are possible, but only if a number of other conditions are met: administrative pluralism is a necessary but insufficient condition. In particular, the consequences of administrative pluralism depend on two things: the degree of closure/indeterminacy of policy networks, and the extent to which communication within, and between, policy networks is open/closed. Thus, any policy network is indeterminate to the extent that its boundaries are vague and the outcomes of its policies are not fixed. When the network's domain is determinate and its policy outcomes are predictable, it has achieved closure. All policy networks are indeterminate to some extent, and the capacity to cope with change and the unintended consequences of change will depend on the openness of communication. As problems succeed problems and problem definitions change the network should become more elaborate – i.e. differentiated – in response to new information about its policies and their consequences. In this ideal world, administrative pluralism realises the advantages of decentralisation and accountability when it has procedures for institutionalising indeterminacy and open communication within networks and between the network and its domain (including the clients of the service).

In reality, however, policy networks practise premature closure. The professions monopolise skills and knowledge. 'The organised few consult the disorganised many', and the professions practise a form of 'exclusionary social closure', through restrictions in labour supply and protection of members from external judgement.[22] Equally, the networks respond to problem succession by 'capture'; that is, multi-networks treaties are negotiated. Thus, Wildavsky argues:

> ... if elements (like departments) are related, so that a change in one mandates a corresponding chain of changes in others, the interaction costs would be prohibitive and the uncertainty boundless. If it were possible, however, to decouple linkages between departments and substitute divisions of labour involving their respective spheres of responsibility, a minimum of interaction and a maximum of unpredictability could thus be maintained.
>
> These treaties internalize externalities by creating even larger departments so that a subject formerly outside of several is now internal to the one.[23]

Indeterminacy and openness of communication are essential attributes of networks if administrative pluralism is not to become the basis for the monopoly of policy-making by circumscribed networks and exclusionary professions. In Luhmann's terms:

> ... a high internal complexity entails allowing alternatives, possibilities of variation, dissent, and conflicts in the system. For that to be possible,

> the structure of the system must be, to a certain degree, indeterminate, contradictory and institutionalised in a flexible way. Against the natural tendency toward simplication and the removal of all uncertainties, it must be kept artificially open and remain under-specified ... the creation of a tolerable indeterminacy in social systems is an achievement and not a mishap.[24]

For example, discussions of accountability in Western Europe tend to focus on the relationship between public sector organisations and elected assemblies, reflecting nineteenth-century preoccupations with representative democracy. In governmental systems with a high degree of internal differentiation, accountability cannot be limited to single organisations but must encompass the network, its relationships *and the policies*: 'the best guarantee of control is designing accountability systems which are appropriate to the policy, rather than specific to the institution'.[25] Equally, considerations of procedural correctness must take second place to network procedures for institutionalising indeterminacy and guaranteeing open communication. These considerations are not the only relevant criteria for evaluating policy networks but, given their near-total omission from the discussion of accountability in SCG, they can safely be emphasised here.

Intergovernmental theory's treatment of territorial interests has been the object of scathing criticism. Tarrow argues that:

> How central governments and their territorial subunits are linked politically is not only a problem of intergovernmental relations but also one of managing the class and interest conflicts of modern societies. No more can intergovernmental relations be separated from political sociology than can the current fiscal crisis be separated from the inner logic of the economic system. Both take territorial forms, but both are ultimately related to conflicts and ideology that emerge from the functional cleavages of a modern society.[26]

Analysis cannot be limited, therefore, to the 'pirouettes of pettyfogging bureaucrats'. In similar vein, Dunleavy argues:

> the organisational and technological logic detected at work in the fragmentation of the extended state is quite general – one might say, socially 'neutral' – it simply cannot be plugged into evidence about substantive social conflicts. It operates from an 'end of ideology' perspective and explicitly asserts the convergence of interest in the modern industrial state between capital, organised labour and government.[27]

Indisputably, too, much intergovernmental theory has focused on bureaucratic politics, ignored distributional questions and, implicitly at least, assumed the convergence of interests. The key question is whether or not these problems are inherent to intergovernmental theory. This work is rooted in the conviction that the approach can be 'plugged into' the analysis of 'substantive social conflicts'. In consequence, the phrase 'territorial politics' has been employed to signal that analysis was not confined to IGR. The definition of 'territorial politics' refers to 'sub-national political organisations' to direct attention to the role of political parties and interest groups as well as government agencies, and the attempt to incorporate a modest historical dimension was the chosen means for relating IGR to social conflicts.

THE PROBLEMS OF HISTORY

Forswearing metaphysical digressions on 'What is history?', the objective of the conferences from which this work emerged was to examine the territorial settlement in place at the end of the 1939–45 war and to explore its subsequent evolution. In effect, an attempt was made to place IGR in its context; to move beyond structures and political/bureaucratic brokerage (or intergovernmental relations) to analyse the range of factors influencing the accommodation of territorial interests (or territorial politics). Especially appropriate is the concept of 'the national government environment' of sub-central government to refer to the changing interventions of the centre and the social, economic and political factors which prompted those interventions. The several country studies reveal both marked changes in the national government environments and an almost embarrassing range of factors prompting such changes. As in the case of the above discussion of the problems of comparison, therefore, the essential first step is to demonstrate that, in spite of the variety, it is possible to generalise about some of the trends in territorial politics.

Three facets of the national government environment proved to be of recurrent importance: the institutional structure, the ideology of central élites and the party system. Rather obviously, the position of local authorities in federal systems such as West Germany differs from that in unitary systems and, for the latter, there is a relevant distinction between codified and uncodified constitutions. Central interventions in local government systems with constitutionally-reserved powers are constrained to a far greater degree than those in systems where local authorities are subject to the doctrines of parliamentary supremacy and *ultra vires*. Of course, legal and institutional factors have been the dominant focus in the comparative study of local government, and criticism of this emphasis is widespread among the contributors to this work. Moreover, as Theo Toonen's analysis of the Netherlands demonstrates, yesterday's doctrine still exercises a pervasive influence. But casting off the detritus of yesteryear does not require a total neglect of institutional factors. As several contributors demonstrate, the centre's relationships with sub-central governments are asymmetric: the centre sets the boundaries to sub-central actions. This power has its roots in the centre's constitutional and legislative resources, and to ignore such factors is to impoverish analysis. Legal-institutional factors do not constitute the whole of comparative analysis but they remain an essential component.

Tarrow distinguishes three types of élite ideology – normative equality, technocratic reformism and distributive welfare – and relates them *inter alia*, to policy impacts.[28] As Yves Mény clearly shows in his contribution, France had *dirigiste* centre–periphery linkages based on an ideology of technocratic reformism for most of the 1960s and 1970s, and the process of democratisation in the 1980s, whatever its other consequences, has enhanced the influence of technocratic elites. Italy had *clientelistic* linkages and an ideology of distributive welfare. However, with the regional reforms of the 1970s, which are explored by Robert Leonardi *et al* in their contribution, traditional linkages were supplanted to some extent by sectoralisation or vertical, function-specific channels of communication and influence. The era of the technocratic élites was dawning, provoking a territorial response in the form of demands by regional politicians both for more effective co-ordination at the centre and for an adequate tax base (to reduce their dependence on central, functional ministries).

Ideology not only distinguishes between the several territorial systems but also, as the change in ideology in response to the world economic recession denotes, it identifies at least one similarity: the phoenix-like revival of nineteenth-century economic liberalism, with its doctrine of the minimal state and its challenge to the inevitability and desirability of welfare expenditures. The UK may provide the most strident example of distributive welfare under threat but it is not an isolated case, as Jens Hesse on the Federal Republic of Germany and Theo Toonen on the Netherlands demonstrate. This change prompts a more important reflection. In spite of a common economic problem and shared ideological preconceptions, the political responses of the several centres have been significantly varied. The UK may have off-loaded to the periphery at the same time as it intensified its efforts to control sub-central governments. But this was not the case with other countries. The Dutch reforms were ostensibly decentralist; the Federal Republic of Germany returned to some extent to the *status quo ante* following federal experimentation with intervention; and both France (after 1981) and Italy created a stronger sub-central tier with the inception of directly elected regional governments. As Jim Sharpe provocatively argues, economic circumstances are all too frequently accorded a determining influence they do not exercise in reality. The economic context imposes constraints and it may even generate imperatives for action, but it does not dictate the *form* of response: political traditions and choices determine courses of action even if, once chosen, they constrain future options.

The primary vehicle for the articulation of both ideologies and alternative courses of action is the political parties. Thus, Jens Hesse's chapter on the Federal Republic of Germany and Yves Mény's on France demonstrate that the changing characteristics of the party system have been of some significance for IGR. In West Germany, the demand for participation from below and competition between the parties for the support of the new interest groups, coupled with long-standing features of the system such as coalition governments and the need for *Bundesrat* consent, have intensified the search for consensus between all major parties for important policies and severely constrained the ability of the federal level to pursue centralising strategies. In sharp contrast is the UK where the tradition of a strong executive fostered unilateral action and the adoption of a command territorial operating code by the centre. In Italy, as Robert Leonardi *et al* show, regionalism was put on the political agenda at the end of the Second World War. But it was always intimately enmeshed in wider party strategies, and the regional reforms of the 1970s were entangled with broader questions of the diffusion of party power.

A similar situation pertained in France throughout most of the Fifth Republic, as Yves Mény makes clear.

But again divergent responses are accompanied by a degree of convergence. A feature of territorial politics from the 1970s onwards has been its *politicisation*. The post-war territorial settlement was rooted in the expansion of the welfare state. Quiescence had its price which was not simply sub-central expenditure growth but also enhanced responsibilities and discretion in the implementation of welfare policies. Recession generated a resource squeeze, constrained discretion and, in the guise of privatisation, eroded responsibilities. Whether in the form of confrontation (UK), demands for an elastic fiscal base and improved central co-ordination (Italy and France), the conflict between 'production-oriented' and 'reproductive' interests (FRG) or the increasingly fragile foundation of 'pillarisation' (the Netherlands), recession demonstrated that the post-war territorial settlement was conditional. For most of the post-war period quiescence was taken for granted, and compliance grounded in incentive (grant) systems was virtually a standard operating procedure in the functional politics of the welfare societies. Recession was an unwelcome reminder that the politics of territory were not inevitably outside the centre but could become an integral part of the politics of the centre to a degree which, for those who had forgotten their history, could not be anticipated.

Moreover, the resurgence of the territorial politics was not solely a product of failure – of expenditure cuts – but also of success – of the dominance of functional politics. As Jim Sharpe has argued:

> the decentralised trends in the politics of the West are, paradoxically, also a product of the centralisation of society and the state machine. That is to say, they are a *reaction* to centralisation and not a mere epiphenomenon of it ... we must acknowledge the possibility that the very centralising socio-economic forces generate a *political* reaction among the putatively integrated.[29]

The expanded welfare state contained the seeds of its own politicisation, recession adding urgency and immediacy to the political reaction.

In the process of politicisation, territorial and other interests became enmeshed and, at times, seemed indistinguishable. Whether the terminology is that of sectoral cleavages (UK) or production-oriented *vs* reproductive interests (FGR), sub-central governments became one of the principal areas within which social conflicts were played out. The differentiation of government was matched by a social differentiation in which functional cleavages were cross-cut by race, gender, nationality and sectoral location (or dependence on government-provided services).

The consequences of such differentiation were varied. Conflict may force governments to be responsive but it can also be diffused through multiple jurisdictions and thereby attenuated, thus removing the imperatives for change. Multiplying the points of access may serve to increase the opportunities for participation but it also increases, as the case of France demonstrates, the opportunities for interests which are already privileged. Cross-cutting cleavages may produce unexpected coalitions of interest and enhance the capacity of the powerless to influence government policy, but they may also erode the traditional bases of support for working-class parties. Indeed, a recurrent problem for contributors was *not* the incorporation of substantive social conflicts into the analysis of territorial politics but the identification of distinct territorial interests among the range of social conflicts for which sub-central government provided the arena. However, before turning to the problems posed by the historical analysis, there is one point worth repeating, even though it should be obvious from the foregoing discussion. It is imperative that the study of any system of IGR be wedded to an analysis of the larger political system of which it forms part: as Theo Toonen clearly argues, when the two are divorced, the analysis of IGR becomes thoroughly misleading.

Surveying developments in the post-war period may have distinct advantages but it also adds a number of problems to the analysis of territorial politics; in particular, the limits to historiography, the boundaries to historical analysis and the identification of territorial interests.

With perhaps deliberate arrogance, Bulpitt warns political scientists about relying on historians to tackle a political science problem: 'Historians are notorious for their disagreements, their lack of conceptual skill, and their liking for story-telling rather than systematic analysis'. Furthermore, 'most historians are not interested in the problems and demands of political scientists'. In consequence, the available literature 'is not only vast but also unhelpful as well', a problem compounded by problems of abstraction and interpretation.[30] These comments are not directed against historians, nor are they designed to scare political science off the study of history. They are, instead, a warning about the amount of work and the difficulties involved; the historical perspective in political science cannot be reduced to 'the usual quick glance through the basic texts'. Such warnings are as apposite for the present exercise in contemporary history as they are for Bulpitt's own, regrettably unpublished, essay on political integration in Tudor and Stuart England. The most important problem of history for the political scientist concerns interpretation. The need for systematic and explicit theory is not met by resorting to historical analysis; if anything, it is intensified. The range of competing, if not mutually exclusive, interpretations remain vast, and without a theoretical framework, the journey into history becomes a fishing expedition, netting an eclectic range of ill-defined variables with indeterminate relations to each other. The contributors encountered this problem most acutely in the form of delimiting the boundaries to the analysis.

The choice of the post-war period fell foul of 'the seamless web of history'. The starting date of 1945 was an arbitrary choice and it would be unwise to ignore the formative influences originating in the nineteenth century. Traditions of political accommodation (in the Netherlands), the specific nature of state formation (in Italy and Germany), the process of social integration (in France), and almost everywhere in Western Europe the twin processes of industrialisation and urbanisation, the growth of central bureaucracy, the emergence of the political ideology of egalitarianism, and the impact of war (especially on states with contested boundaries) are all features of the socio-economic context which exerted centripetal pressure in the development of the modern nation state. Regional distinctiveness, the conflict between town and country and political cleavages in multifarious guises have all exerted centrifugal pressures. Factors such as size, topography and the efficacy of the central state are merely some of the omissions from this illustrative listing. Indeed, the profusion of approaches clustered under the label 'centre–periphery relations' reflects (in part) the need to select and thereby simplify the analysis of the formation of the modern nation state. We, too, have selected and in so doing extended but not fundamentally altered, the preoccupations of intergovernmental theory; a visual filter which directed attention towards governmental institutions, political and administrative brokerage and the ideology of central élites. The analysis of the context, of substantive social conflicts and of trends represent an extension of intergovernmental theory. But they are merely extensions to that theory's particular concerns. Thus, the changing ideology of central élites is examined only in so far as it affects political and administrative brokerage, while the analysis of the social, economic and political context isolates only those features with a direct bearing on IGR. Our claim to have incorporated a limited historical dimension to the analysis of IGR has to be viewed, therefore, with some caution, although not with outright scepticism. We have identified variables of central importance to the analysis of territorial politics but the analysis remains perhaps too tightly focused: a preoccupation with middle-range theorising may betray the inherent fissiparousness of historical enquiry.

In spite of our (relatively) precise definition of the field of study, the concept of 'territorial interests' has proved to be as elusive as it was essential. Social conflicts are played out within territory and across territories but are only rarely confined to distinct territorial interests. The following contributions describe and analyse the inter-relationship between interests and territory without resolving the problem. As a way of both recognising the importance of the concept for our deliberations and stimulating further enquiry, Jim Sharpe was asked to write a concluding essay on some of the wider theoretical and methodological implications of the work undertaken in this collection. Sharpe demonstrates that political science has ignored the concept and, drawing on the country studies, shows that the omission is a mistake. It is to be hoped that this essay in particular, and the collection as a whole, play a part in restoring territory to its rightful place on the agenda of political science research.

This discussion of the problems of adding a historical dimension to the study of IGR should not be viewed as an exculpatory exercise, designed to divert criticism. In seeking to establish the case for incorporating history to the study of IGR, there is no need to claim that our version of the merger realises its full potential, merely to demonstrate that there is potential. It is with a discussion of the future potential (and expansion) of the field that we conclude this introduction.

CONCLUSIONS

To provide a summary of an introduction which is itself a summary of objectives, themes and problems is to carry redundancy to intolerable lengths. It will be more productive to identify avenues of future research which address the problems of the field.

In the first place, although the resource squeeze on the sub-central governments has been ubiquitous, there have been no studies which go beyond general description and compare the consequences of the varied central strategies for the range of affected interests associated with a specific welfare service. Such a precise focus on a single service should permit specific answers to the following kinds of issues:

- The degree to which problems are factorised: for instance by-passing elected authorities and resorting to indirect administration.
- The degree to which policy networks are professionalised.
- The extent to which networks have established client groups which can be mobilised in their defence.
- The extent to which heterogeneous interests are integrated within the networks and insulated from other networks.
- The degree of dependence of central actors on other network members not only for the implementation of policy but also for initiatives.
- The extent to which accommodation within and between networks is a 'limit to rationality'.
- The degree of congruence between central élite ideology – its operating code – and the differentiated and disaggregated polity.
- The extent to which networks constrain innovation at all levels of government.
- The ways in which networks (1) exclude disadvantaged groups and (2) reduce the scope for participation by all citizens.
- The distributional consequences of networks, more especially the extent to which the fragmentation and professionalisation of service delivery systems sustain an inequitable distribution of resources/ services; and, under conditions of resource squeeze, accentuate such inequalities.

To a degree denied to a general survey of trends, this kind of research will encourage the exploration of substantive social conflicts through its concern with the distributional consequences of resource squeeze.

To complement this detailed study, a second fruitful avenue of research would be the changing territorial operating code and strategies of the centre. In differentiated and disaggregated polities attention is understandably directed towards sub-central governments. But it is as important to confront the question of 'What is the centre?' To rephrase the question, the definition and location of 'the centre(s)', its operating code and strategies of intervention have changed over time. An adequate analysis of the national government environment requires an explanation of how and why it has changed. Such a study could take many forms, ranging from a historical survey of the growth and development of the central bureaucracy to 'biographies' of particular ministries. As Bulpitt argues there has been a reluctance 'to examine seriously the resources, intentions and operations of the central authorities' and yet it is 'the only way to provide a macro-development study of territorial politics'.[31] To examine the centre is to repair a gap in our knowledge and to prompt the cohabitation of history and political science, thereby enhancing the prospects of a fruitful marriage.

INTRODUCTION   19

Quite obviously, there is a multiplicity of research avenues worthy of exploration. Our objective is not to constrain the choice of topics but to demonstrate that much research of value remains to be undertaken; to suggest that the twin foci of interests and 'the centre' are particularly promising avenues; and to proselytise for a historical perspective. Any collection should provide some essays which are interesting and/or useful to particular readers. In this case, we are tempting fate with our ambition and claiming that, taken together, the contributions demonstrate the utility of a focus on territorial politics and of adding a historical dimension to the study of IGR.

NOTES

1. J. G. Bulpitt, *Territory and Power in the United Kingdom* (Manchester: Manchester University Press, 1983), Chapters 1 and 2.
2. See also S. Tarrow, *Between Center and Periphery* (New Haven: Yale University Press, 1977), pp. 18–35.
3. For a review and criticism of the existing literature see R. A. W. Rhodes, *Control and Power in Central-Local Government Relations* (Farnborough: Gower, 1981), Chapter 4.
4. See, for example, B. C. Smith, 'Measuring Decentralisation' in G. W. Jones, (ed.) *New Approaches to the Study of Central-Local Government Relationships* (Farnborough: Gower, 1980), pp. 137–51, and the debate between E. Page and G. W. Jones and J. D. Stewart in *Local Government Studies*, Vol. 8, No. 4, 1982, pp. 21–42; Vol. 8, No. 5, 1982, pp. 10–15; Vol. 9, No. 1, 1983, pp. 18–21.
5. See T. Nairn, *The Break-Up of Britain* (London: NLR, 1977); M. Hechter, *Internal Colonialism: The Celtic Fringe in British National Development 1536–1966* (London: Routledge & Kegan Paul, 1975). For criticisms see E. Page, 'Michael Hechter's 'Internal Colonial' Thesis: Some Theoretical and Methodological Problems', *European Journal of Political Research*, Vol. 6, 1978, pp. 297–317.
6. See, for example, S. Rokkan, *Citizens, Elections, Parties* (New York: McKay, 1970); S. Rokkan, 'Dimensions of State Formation and Nation Building: A Possible Paradigm for Research on Variations within Europe' in C. Tilley (ed.), *The Formation of National States in Western Europe* (Princeton, N.J.: Princeton University Press, 1975), pp. 562–600; S. Rokkan and D. W. Urwin (eds.), *The Politics of Territorial Identity* (London: Sage, 1982); S. Rokkan and D. W. Urwin, *Economy, Territory, Identity: The Politics of the European Peripheries* (London: Sage, 1983).
7. For example, M. Crozier and J. C. Thoenig, 'The Regulation of Complex Organised Systems', *Administrative Science Quarterly*, Vol. 21, 1976, pp. 547–70; Tarrow; R. A. W. Rhodes, *The National World of Local Government* (London: Allen & Unwin, 1986).
8. Bulpitt, p. 1.
9. Bulpitt, pp. 53, 54.
10. R. A. W. Rhodes, '"Power-Dependence" Theories of Central-Local Relations: a critical re-assessment' in M. J. Goldsmith (ed.), *New Research in Central Local Relations* (Farnborough: Gower, 1986), pp. 1–33.
11. Bulpitt, p. 54, and Urwin, 'The Price of a Kingdom: Territory, Identity and the Centre-Periphery Dimension in Western Europe', in Yves Mény and Vincent Wright, *Centre-Periphery Relations in Western Europe* (London; Allen & Unwin, 1985), pp. 151–70.
12. Bulpitt, pp. 54–5.
13. The editors are currently working on a collection of historical essays, many of which were presented at the same two conferences as the papers in this collection.
14. K. Newton, *Balancing the Books* (London: Sage, 1980), pp. 12–13.
15. Notable exceptions include Rokkan and Urwin; Tarrow; S. Tarrow, P. J. Katzenstein and L. Graziano (eds.), *Territorial Politics in Industrial Nations* (London: Praeger, 1978); and L. J. Sharpe (ed.), *Decentralist Trends in Western Europe* (London: Sage, 1979).
16. L. J. Sharpe, 'Is There a Fiscal Crisis in Western European Local Government? A First

Appraisal' in L. J. Sharpe (ed.), *The Local Fiscal Crisis in Western Europe: Myths and Realities* (London: Sage, 1981), p. 24.
17. P. Dunleavy, 'Professions and Policy Change: Notes Towards a Model of Ideological Corporatism', *Public Administration Bulletin*, No. 36, August 1981, pp. 3–16; and Rhodes, *The National World of Local Government* (London, Allen & Unwin, 1981).
18. N. Luhmann, *The Differentiation of Society* (New York: Columbia University Press, 1982), pp. xv and 353–5.
19. V. Wright, 'Regionalization under the French Fifth Republic: the Triumph of the Functional Approach', in Sharpe, *Decentralist Trends in Western Democracies*, pp. 194–234.
20. D. S. Elgin and R. A. Bushnell, 'The Limits to Complexity: Are Bureaucracies Becoming Unmanageable?', *The Futurist*, December 1977, pp. 327–49.
21. P. Dunleavy, 'Quasi-government sector professionalism' in A. Barker (ed.), *Quangos in Britain* (London: Macmillan, 1982), pp. 185–6.
22. F. Parkin, *Marxism and Class Theory: A Bourgeois Critique* (London: Tavistock, 1979), pp. 54–8.
23. A. Wildavsky, *The Art and Craft of Policy Analysis* (London: Macmillan, 1979), p. 74.
24. Luhmann, pp. 147–8.
25. V. Gray, 'Accountability in the Policy Process: An Alternative Perspective' in S. Greer, R. D. Hedland and J. L. Gibson (eds.), *Accountability in Urban Society* (London: Sage, 1978), p. 170.
26. S. Tarrow, 'Introduction' in Tarrow, Katzenstein and Graziano, pp. 1–2.
27. P. Dunleavy, 'The Limits to Local Government' in M. Boddy, and C. Fudge (eds.), *Local Socialism?* (London: Macmillan, 1984), p. 60.
28. Tarrow, *Between Center* ... pp. 33–5.
29. Sharpe, *Decentralist Trends* ... p. 20.
30. J. G. Bulpitt, *The Problem of 'The Northern Parts': Territorial Integration in Tudor and Stuart England* (Coventry: University of Warwick, Department of Politics, Working Paper No. 6, 1975), pp. 33, 41.
31. Bulpitt, *Territory and Power*, pp. 57–60.

Territorial Politics in the United Kingdom: The Politics of Change, Conflict and Contradiction

R.A.W. Rhodes

Britain is said to be 'the home of local government'; a phrase which means presumably that local authorities are directly elected, large, multi-functional, wealthy — with the freedom to raise a large proportion of their large income from local taxes (the rates) — and with a considerable degree of discretion in decision-making on local services. Yet this bliss has been shattered by dissent for nearly a decade. Condemnations of centralising government policy abound. Prognostications about the demise of local government are as frequent as critiques of excessive centralisation. Quite obviously, the central questions to be addressed in this article are what has changed and why.

The questions might seem straightforward, but the answers are not easy. It is wholly inadequate to lay the blame at the feet of the post-1979 Conservative governments and their commitment to social-market liberalism and the minimal state. Current conflicts have long-standing roots, and yet little or no attention has been paid to the origins of the system being transformed. It is essential to specify the pre-existing characteristics of central–local relations before changes can be identified with accuracy. And the origins of the system lie in the processes of territorial integration in Tudor and Stuart England. By the end of the seventeenth century, 'only a partially integrated polity had been achieved, a hybrid polity where intense local loyalties existed side by side with the acceptance of central authority and a central culture'. By the twentieth century, this division between court and country had been elaborated into a Dual Polity with three defining characteristics: a structural dichotomy between centre and periphery; a distinction between high and low politics; and a territorial operating code — or rules of statecraft — which emphasised central autonomy to pursue matters of high politics. Consequently, 'local politics have remained more separate from national politics than in almost any other country'.[1] To focus, for example, on government control of local expenditure is to adopt, therefore, a historically parochial perspective on the question of what has changed. More fundamentally, the foundations of the Dual Polity have been riven. Consequently the bounds of analysis have to be extended beyond central–local financial relations to the structure of political relationships: to the changing nature of the Dual Polity.

It is also necessary to extend analysis beyond the study of central–local government relations to encompass sub-central government (SCG) in its entirety. A central characteristic of British government since 1945 has been the development of differentiated or fragmented service delivery systems. The centre has had 'hands-off' controls over a multiform institutional structure at a time when, under the exigencies of, *inter alia*, economic decline, its operating code has demanded both autonomy in economic management and intervention to gain that autonomy. The objective of this analysis is to provide a sketch of developments in SCG in the post-war period and explain why the Dual Polity in all its institutional guises has collapsed.

Given that 'SCG' is yet another addition to the 'alphabet soup' of British governmental acronyms it would be as well to define it. Figure 1 presents a summary of the descriptive terminology to be employed.

FIGURE 1   THE SCOPE OF SUB-CENTRAL GOVERNMENT

```
                    SUB-CENTRAL GOVERNMENT
        ┌──────────────────────┴──────────────────────┐
   Territorial representation         Intergovernmental relations
                                    ┌───────────────┴───────────────┐
                                Central–local        Inter-organisational
                                 relations                relations
```

Territorial representation refers to the representation of ethnic-territorial units (i.e. Scotland, Wales and Northern Ireland) at the centre and the activities of territorially based political organisations. *Intergovernmental relations* (IGR) refers to the relations between central political institutions and all forms of governmental organisations beyond the centre. It is sub-divided into *central–local relations* or the links between central departments and local authorities and *inter-organisational relations* which encompasses all other types of sub-central government (i.e. nationalised industries, the national health service, the 'regional' organisation of central departments and non-departmental public bodies). In short, SCG refers to that arena of political activity concerned with the relations between central political institutions in the capital city and those sub-central political organisations and governmental bodies within the accepted boundaries of the state: a broad definition that immediately makes clear that British SCG is complex and not limited to local government (supra, p. 2). The importance of this extension will become obvious from the subsequent discussion.

The remainder of this contribution provides, first, a framework for analysing territorial politics; second, a description of developments in SCG between 1945 and 1985; and, third, an explanation of the major changes. It concludes with a summary which emphasises that Britain is best characterised as a differentiated polity, rather than a unitary state, in which interdependence and ambiguous and confused relationships foster the co-existence of both fragmentation and centralisation.[2]

CONCEPTS AND PROCESSES

In order to analyse the transformations in SCG, it is necessary to distinguish clearly between the macro, meso and micro-levels of analysis. At the macro-level, the politics of SCG have become part and parcel of the politics of Westminster and Whitehall and, without providing a comprehensive account of changes in British government, it is necessary to explore the *national government environment* or central government institutions and their socio-economic environment as they affect SCG.[3] At the meso-level, it is necessary to look for regularities in the relationships between the range of sub-central bodies.

The concept of *policy networks*, or complexes of organisations connected to each other by resource dependencies and distinguished from other complexes by breaks in the structure of resource dependencies[4] is a useful tool for ordering the profusion of links. Finally, at the micro-level, it is necessary to explore the behaviour of particular *actors*, be they individuals or organisations. Their behaviour is not simply a product of their environment: they can shape that environment to greater or lesser degree.

It will not be possible either to describe all species of SCG or to explore all three levels of analysis. The emphasis falls on the national government environment and on policy networks in general, and, to be more specific, on the processes of change within the national government environment and their (reciprocal) effects on policy networks. For present purposes, six processes are of key importance: instability of the external support system, decline of the mixed economy, increasing functional differentiation (coupled with professionalisation) within the welfare state, the oscillating conflicts between functional and territorial politics, political differentiation and change *and* continuity within the British political tradition (most notably in the territorial operating code of central élites). Each process is introduced *seriatim*: their inter-relationships are described in the next section.

The External Support System

Bulpitt, from whom this phrase is taken, argues not only that 'any centre will attempt to minimise the impact of external forces on domestic politics' but that the British centre has been unable to maintain a stable external support system. Thus, by the 1970s, Britain had lost its capacity for independent military action (e.g. Suez), its old alliances had lost their effectiveness (e.g. the Commonwealth, the 'special relationship' with the USA), and sovereignty was qualified by membership of international organisations (e.g. EEC, NATO) and by dependence on the international economy (e.g. IMF, OPEC).[5] Britain was an 'open polity', vulnerable economically and politically to external forces. Nowhere was this more evident than in the management of the economy.

The Decline of the Mixed Economy

As the era of 'stop-go' gave way to 'stagflation', the decline of the British economy became an inescapable fact. Domestic economic weakness was exacerbated by an unstable external support system, giving rise to 'de-industrialisation' or the lack of 'an efficient manufacturing sector' which, currently as well as potentially, not only satisifies the demands of consumers at home but is also able to sell enough of its products abroad to pay for the nation's requirements.[6] The economic problem – 'the British disease' – was variously compounded of an aged and ageing industrial base, a declining productive (especially manufacturing) capacity, a lack of international competitiveness, high inflation, high unemployment, an unreformed pay bargaining structure, a large and expanding public sector and, in consequence of some combination of these factors, a low growth rate. Side-stepping the thorny question of the relative importance of these factors, the fact of economic decline had the effect of translating the 'problems' of SCG into an epiphenomenon of national economic management. The centre has 'off-loaded to the periphery'; that is, it has sought to resolve its economic problems initially by modernising the institutions of SCG to make them more efficient and subsequently by directly regulating their expenditure. The sheer size of SCG as a proportion of public expenditure (and employment) made it vulnerable to retrenchment.

Differentiation, Professionalisation and the Welfare State

SCG is large because it was the prime vehicle for building the welfare state up to the 1970s. Central departments are, for the most part, non-executant agencies: they depend on SCG for the implementation of policy. This simple fact has had three important consequences. First, it created interdependence in the form of vertical, function specific links between centre and locality. Second, the expansion of the welfare functions of SCG was not matched by an expansion of its resources. Grants-in-aid financed the expansion and, with the onset of economic decline, the centre had to hand the means for regulating SCG expenditure. It could cut grants, confident that SCG lacked adequate resources of its own.[7] Finally, local government was not the sole beneficiary of the expansion. The postwar years were the era of 'ad-hocracy' as numerous specialised, even uni-functional, invariably non-elected public bodies were created (e.g. New Town Development Corporations, Manpower Services Commission, National Health Service, Regional Water Authorities). With the increasing size of government and the ever broader range of its interventions, functional differentiation became the order of the day: growth by elaboration and specialisation.

Allied to the process of differentiation is that of professionalisation. The specialists in the several policy areas became institutionalised in the structure of government. The professions became key actors in the function specific policy networks. Dependence on expertise fostered functional differentiation and the employment (even creation) of professions which, in turn, accelerated the fragmentation (and complexity) of government.

In short, policy-making in British government became characterised by professional-bureaucratic complexes based on specific functions: a series of vertical coalitions or policy networks in which central dependence on SCG for the implementation of policy was complemented by SCG's dependence on the centre for financial resources. Fragmentation between functions within the centre went hand in hand with centralisation within policy networks; there was no one centre but the 'centreless society' or the differentiated polity with multiple centres wherein fragmentation and centralisation co-exist.[8]

Functional v. Territorial Politics

The emergence of policy networks with the growth of the welfare state could be interpreted as the triumph of functional over territorial politics. Thus, channels of communication between centre and locality were not based on territorial representation but on professional-bureaucratic contacts within policy networks. The politics of service provision, of a centrally dispensed territorial justice, rivalled the politics of place; uniform standards challenged local variety. But the pre-eminence of functional politics created tensions.

As Sharpe[9] argues, the decentralist trends in Western Europe are a political reaction against the integration imposed by the centralising policy networks. Nor is the tension between functional and territorial politics visible only in ethnic nationalism and the devolution debate. On a smaller scale, it is re-enacted within local authorities in the tension between local needs and best professional practice and within the policy networks between the technocratic and topocratic professions.[10] It became most apparent with the politicisation of SCG during the 1970s.

For the bulk of the post-war period, central and local political élites were insulated from each other. Local parties were poorly organised and, in many local authorities, virtually non-existent: dominated by 'a culture of *apolitisme*'. The 'dual polity' prevailed wherein the centre sought to distance itself from SCG and 'Low Politics' (e.g. public health) in order to enhance its autonomy in matters of 'High Politics' (e.g. foreign policy).[11]

However, the modernisation of local government in 1974 politicised local authorities. Nor was this trend evidence of the nationalisation of local parties. They may wear national labels but, as events since 1979 demonstrate, that is no guarantee of compliance. If recalcitrance has been a feature of local Labour parties, it has not been unique to them and many Conservative-controlled councils have forsworn reticence for an overtly critical stance towards the national party's policy. The search for control by the Thatcher governments has provided a considerable jolt in the arm to territorial politics. The perceived effect of control was the erosion of local services. The consequent protests, the search for redress against functional politics, provide yet one more illustration of the oscillating relationship between territorial and functional politics. If the latter has been pre-eminent for the bulk of the post-war period, the former has never been extinguished and its persistent flickerings serve as a reminder that questions of territorial structure lie at the heart of the modern nation state.

Political Differentiation

The fluctuating fortunes of territorial representation are one manifestation of a wider phenomenon; increased political differentiation. Class may remain the basis of British politics, but it is not unchallenged. Language and culture, nationalism, religion and race have all become increasingly salient bases of political allegiance. Partisan de-alignment characterised British politics; that is, electoral allegiance to the two main parties declined, votes were decreasingly cast on class lines, turnout declined and the electorate became more volatile.[12]

To further complicate the picture, new bases of political differentiation emerged: the policy networks generated sectoral cleavages within localities. In brief, services provided by policy networks created client groups with a vested interest in maintaining those services. These clients can be mobilised to resist threatened reductions in expenditure or to highlight gaps in service provision. For example, the protests of middle-class parents at both the reduction in places at universities and the increasing cost of such education have been a constraint on attempts to restrain expenditure on higher education. Conversely, sectoral support for the compulsory sale of council houses has facilitated government action to the discomfort of Labour opposition.

Policy networks can determine the framework within which any debate about their services takes place and can shape local political conflicts. Governmental fragmentation generates social and political fragmentation which both cuts across traditional class allegiances, creating alliances based on mutual dependence on government services, and also reinforces the advantages of privileged groups, providing additional points of access.[13]

Continuity within the British Political Tradition

Without denying that there have been some significant changes, none the less stable elements within the British political tradition have limited the capacity of the centre to respond to the changes touched upon above. Two features of this tradition can be mentioned only in passing. First, Britain is a unitary state wherein 'the power to delegate or revoke delegated power remains in the hands of the central authority'.[14] Parliamentary sovereignty is the cardinal feature of British politics. Second, the two-party system concerned pre-eminently with functional politics serves to integrate the constituent territorial units of the UK. The main concern of this section is with the ideology of central élites. The very phrase suggests that the main topic to be discussed is the shift from Butskellism to monetarism and social-market liberalism under Margaret Thatcher's government. This shift is important, but has been widely noted[15] and will not be elaborated upon here. In marked contrast, two other features of central élite ideology have been unwisely ignored: the reassertion of executive authority and the centre's territorial operating code.

A central feature of social-market liberalism is its belief in the 'minimalist state'. However, the attempt to bring it into being has provided the clearest assertion for more than a decade of the centre's belief in its right to govern. With the ever-increasing functional differentiation of British government, there has been an ever more fraught tension between the tradition that 'leaders know best' and the centre's dependence on SCG for the delivery of services. If control strategies are adopted, then governments have to confront unintended consequences, recalcitrance, instability and confusion. If compliance is sought, then governments have to confront 'slippage' or the adaptation of policies in the process of implementation. And yet it is precisely such slippage which provides the incentive for unilateral action. This tension has plagued all governments in the post-war period − it induces ministerial 'schizophrenia'.[16]

With the introduction of controls to create the minimalist state, the centre abandoned its traditional 'territorial operating code'. This phrase refers to the 'rules of "statecraft" employed over time by political elites' and the code had taken the form of the 'dual polity' in which the centre was insulated from localities allowed considerable operational autonomy. In its place, Bulpitt[17] detects a populist code; the by-passing of intermediate institutions such as local government in favour of direct links with citizens. Perhaps more significantly, the government has forsaken central autonomy to pursue 'High Politics' for a command, or a bureaucratic hierarchy, model of domestic intervention. Repeated interventions to exact control at the level of the individual local authority bespeak not populism but the foreman and the shop floor. Such a code may be congruent with the exercise of executive authority but it is faulty, representing a failure to appreciate the complexity of the differentiated polity.

These remarks on the processes of change within the national government environment are introductory. The value of such a focus can only be established by surveying the evolution of SCG in the post-war period.

THE DEVELOPMENT OF SCG

Five phases in the development of SCG can be identified. The years 1945–61 were a period of growth in welfare expenditure, grant consolidation and quiescent territorial politics. From 1961 to 1974 was the era of institutional modernisation, intergovernmental bargaining and territorial protest. This was followed, 1974–79, by an era of economic decline in the UK, in which local élites were incorporated into central decision processes and territorial problems were factorised. From 1979 to 1983 was the era of sustained economic recession, repeated central interventions to control sub-central institutions (and expenditure) and a centralisation of territorial politics. From 1983 multiplying contradictions between centralisation, politicisation and deregulation have been witnessed.

1. 1945–61: Growth, Consolidation and Quiescence

Strictly speaking, the years 1945–51 should be treated as the era of post-war reconstruction and the subsequent decade as one of growth. Any such division on economic grounds masks, however, other substantial continuities; with the benefit of hindsight the era is significant for what did not happen. Nationalist parties were conspicuous only for their electoral weakness. The seeds of civil unrest in Northern Ireland were being sown by the Unionists but the crop had yet to be reaped. Conflict between central and local government was spasmodic and had a certain novelty value. If any concern was expressed it focused on the loss of functions by local government and the growing dependence of local authorities on central grant; the perennial cry of centralisation yet again rent the air. None the less the era has a number of features of considerable significance.

First, the period was one of external stability, low unemployment, low inflation and relatively high growth rates. This economic surplus provided the means, while the central élite ideology provided the motive, to create the welfare state, and sub-central government was to be the prime vehicle for the delivery of its services. Central departments in Britain were and have remained non-executant units of government. They have 'hands-off' control of such major services as housing, education and health and welfare. Thus, if the rate of growth was to accelerate in 1961–74, the foundations were dug for that development in the preceding period.

Second, the arrival of government intervention and the welfare state in Northern Ireland sowed further seeds of discontent. Stormont may have intervened reluctantly, the welfare state may have failed to redistribute resources to those in the greatest need, but both posed in acute form questions of relative deprivation[18] long before the Civil Rights Movement. The contrasting levels of affluence of the mainland and the Province, of Protestant and Catholic, coupled with rising expectations about entitlements, served to emphasise the fragility of devolution.

The third major development was the increasing use by government of *ad hoc* agencies, or non-departmental public bodies. The best known of these agencies is, of course, the National Health Service, but the period also brought the removal of public utilities from local government and the creation of the nationalised industries (e.g. gas, electricity).

Fourth, the period saw the increasing prominence of professionals, not as powerful interest groups lobbying government, but as part of the structure of government; they became institutionalised. The long-established professions in local government emerged to prominence in the 1930s. This was the period when the accountants, lawyers, engineers and public health inspectors consolidated their position and such 'newcomers' as teachers, social workers and planners arrived. For the most part, career advancement was solely within the public sector and professional work organisation and the departments of local government were identical. The expansion of the welfare state went hand in hand with the extension of professional influence and the emergence of functional politics.[19]

Fifth, central funding of local services was 'consolidated'. During the 1930s some 35 per cent of central funding was in the form of general grants: that is, moneys were *not* assigned to specific services. By 1961, this figure had risen to 68 per cent. And yet, one would expect the central departments with responsibilities for particular services to resist such an erosion of their influence – and indeed the (then) Ministry of Education did oppose such consolidation in 1929. Central resistance to block or general grants evaporated in this period because 'other constraints on local government emerged in Britain alongside the growth in the grant system, so that specific grants became less necessary in the influencing of local service delivery'. These constraints include the development of the legal framework of services; the vertical coalition of professionals in central and local government with shared values about (and responses to) service delivery; and finally the creation of vested interests by the specific grant so that producers *and* consumers support service expenditure.[20]

Finally, the period is distinguished by *'apolitisme'*. There are a number of strands to this argument. First, party colonisation of sub-central politics was incomplete and, for example, the numerous *ad hoc* agencies were subject to otiose forms of accountability. Second, a substantial proportion of local councils were controlled by 'Independents' not the political parties. Third, the incursion of party politics into local government was resented and resisted – after all, 'there's only one way to build a road'. Fourth, even when the parties did control local authorities, this control could be purely nominal with little or no impact over and beyond the election. Decision-making was the preserve of committee chairmen and chief officers. Above all, IGR were characterised by professional-bureaucratic brokerage and the relative weakness of political linkages between centre and periphery (for a more detailed discussion see Bulpitt).[21]

In the light of subsequent developments, this was the era when that hackneyed dog did not bark. Scottish interests were accommodated by both the growth of expenditure and the gradual expansion of the functions of the Scottish Office.[22] The same central strategy of accommodation through economic and administrative growth was applied to local authorities and the Northern Ireland Office, the only exception being Wales. To a significant degree, sub-central and central interests were united in the development of the welfare state and the pursuit of economic growth. The consolidation of functional politics was founded on consensus and quiescence, *and* served the interests of the centre.

2. 1961–74: Modernisation, Intergovernmental Bargaining and Territorial Protest

'The white heat of the technological revolution' did not herald the modernisation of the British economy — that lies with the Macmillan–Maudling experiment with planned growth from 1961 — but it provides the era with one of its more potent symbols. Along with the introduction of national planning, regional planning and a new national budgetary process came a more interventionist style of government and extensive reform of the machinery of government at every level. Britain was to be modernised.

And yet consultation and bargaining was the normal style of intergovernmental relations. For most of the period local service spending was buoyant. Central governments of both parties kept an eye out for the electorally damaging implications of any slippage by local government in areas of key importance. For example, road construction, slum clearance and rehousing were the major public concerns for most of the 1950s and 1960s, as was the reorganisation of secondary schooling from the mid-1960s until the mid-1970s. A whole series of expectations about reasonable consensual dealings between Whitehall and local councils were embodied in the concept of 'partnership'. Ministers often went out of their way to choose modes of implementing policy that maximised voluntary local authority co-operation.

The bargaining phase lasted throughout the Heath government's period of office, despite some selective attempts by the Conservatives to develop more stringent controls. Initially, the government's policy for the nationalised industries rejected intervention: 'lame ducks' were 'to go to the wall' and profitable industries were to be 'hived off'. The policy was not long-lived nor did it have a major impact on SCG. With somewhat greater determination, the government forced through changes in council housing finances against strong resistance (including the attempt by the Labour council at Clay Cross in 1972 to refuse implementation of the rent increases imposed). But elsewhere the government was cautious. Sales of council housing were successfully obstructed by all Labour councils. And although a full-scale reorganisation of local authorities was put through against much opposition from councils destined to lose many of their powers, the Conservative government adopted a two-tier system which both protected party interests and was more popular with existing councillors and officers than previous proposals for unitary authorities. Moreover, having decided on the principles of reform, the government was prepared to bargain over such 'details' as the allocation of the planning function. Nowhere is evidence of bargaining clearer than in the determination of the level and allocation of central grants. Through their national representative organisations, local authorities were able to gain small but significant changes in the total, rate of growth and distribution of central grant. For local government, therefore, the period embraces both centrally initiated reform and consultation.

Modernisation intensified conflict between national and sub-central units of government. Central intervention provoked confrontations in the fields of education and housing. But given the scale of institutional change, the increase was modest. In part, protest was limited by the pre-existing structure of IGR. Local access to national political élites was relatively weak whereas professional actors were key advocates of the reform.[23] Perhaps most important, the centre 'factorised' the problem of reform. Thus, there were separate reorganisations of local government in Scotland, England and Wales, London and Northern Ireland; of particular functions (e.g. water, health); and of the centre's decentralised arms (e.g. regional economic planning councils, Welsh Office). At no time was the reform of IGR comprehensively reviewed. Key aspects of the system were ignored altogether (e.g. finance). In the case of water and health, reform simply served to reinforce professional dominance. In the case of the nationalised industries, the post-1961 era of commercial freedom had returned many to profitability. The price controls of the Heath government were to reduce their finances to chaos. The strategy of factorised modernisation had few obvious economic benefits, although it encouraged a fragmented response and dampened the level of protest.

If during this period functional politics continued to fragment, there were also major conflicts in the arena of territorial representation. In the late 1960s, the introduction of direct rule in Northern Ireland and the rise of Scottish and Welsh nationalism represented a far greater threat to the consensus which had governed territorial politics in Britain.

Explanations of the rise of nationalism abound. The factors which caused the resurgent electoral performance of the SNP include: the decline of the UK economy, loss of confidence in British government, institutional weakness, the relatively greater economic decline of the regions, cultural differences, nationalist feeling, and specific issues or grievances – for example, North Sea oil, membership of the EEC. There is corresponding disagreement on the importance of these factors.

Nationalism is not a phenomenon of the 1960s and early 1970s. It is a persistent feature of the British political landscape reinforced by separate education, legal, religious and governmental systems. Nor can such popular expressions of nationalism as a separate international football team, a national flag and a national anthem be omitted from this list of distinctive characteristics. Although it is a long-standing phenomenon, the strength of Welsh and Scottish nationalism can be overstated. As Rose[24] has pointed out, 'By fighting elections, Nationalists register the weakness of their support in their own nation'. The 'problem' to be explained is not, therefore, the rise of nationalism but why the centre in this period took the challenge so seriously. In other words, the issue to be explored is the constraints upon and the weakness of central government. In addition, the increasing salience of nationalism cannot be divorced from the increasing salience of other social and political cleavages. The combined effects of central weakness and political differentiation were to become marked from 1974 onwards.

Above all, the prominence of nationalism in this period should not obscure the substantial continuities in IGR. The policy networks remained paramount in the expansion of welfare state services; the centre remained politically insulated from local élites; and in Urwin's [25] phrases 'tolerance and indifference'; 'the concern to accommodate demands within the prevailing structure'; and 'an *ad hoc* attempt to resolve a specific complaint or demand' continued to characterise central attitudes and actions.

The key feature of intergovernmental relations in this period is instability, not crisis. Economic pressures were mounting, the central strategy of institutional modernisation was bearing no obvious fruit and the incidence of conflict and protest was increasing. The onset of economic decline was to alter the picture markedly.

3. 1974–79: Economic Decline, Incorporation and Territorial Factorising

Three factors are particularly important for understanding developments in this period. First, the economic decline of the UK was accelerated by an unstable external support system. This instability was economic – the escalation of world commodity prices, most notably oil prices – and political – UK dependence on international bodies and corresponding inability to take independent action. Confronted by massive inflation, the government had to seek a substantial loan from the IMF which required drastic cuts in public expenditure. Second, the reorganisation of local government stimulated the spread of party politics and the virtual demise of the 'Independent'. The consequent politicisation of local politics was to spread to intergovernmental relations. Third, the 1974 election resulted in a Labour government with a majority of four, and these knife-edge situations generated imperatives to negotiate with minority parties to preserve a working majority. This conjunction of external economic disruption and political fragility called forth central strategies of incorporation for English local authorities and factorising for nationalist political demands.

There is clear evidence that the Labour government did make a sustained effort to introduce a kind of top-level, overall 'corporatism' into its dealings with local government. Its essential innovation was to try to incorporate the powerful local authority associations (and their joint bodies) into a sort of 'social contract' about local government spending. For the first time, Whitehall set up a forum in which to discuss the long-run future of local spending with the local authority associations. This Consultative Council on Local Government Finance (CCLGF) was remarkable also in bringing the Treasury and local authority representatives into face-to-face contact for the first time, and in explicitly integrating the planning of local spending into the Public Expenditure Survey system. The government's hope was that by involving the local authority associations in policy-making affecting local government they would be able to persuade them of the 'realities' of the economic situation, and thus enlist them as allies in the battle to keep down the growth of local spending. In the Treasury's view, the objective of the CCLGF was 'effective control' and if it was not achieved, 'other measures would have to be considered'.[26]

How effective the CCLGF was in meeting this aim is difficult to say. Many of its members argue that it was successful in getting the local authority associations to persuade their members to behave with restraint. But, of course, there were other forces working in the same direction: for example, cash limits on central grants to local authorities; the swing in mid-term local elections to Conservative-controlled authorities committed to expenditure restraint. Whatever else it accomplished, however, the Council did help to shift influence within local government away from service-orientated councillors and officers (for example, the education policy community) and towards local politicians and finance directors more concerned with 'corporate planning', increased efficiency and financial soundness. In effect, the Council was an attempt by Whitehall to build up the influence of the national community of local government, in order that it would be better able to control the rest of the local government system in return for consultation and a direct voice in future planning.[27]

The importance of aggregating interests was matched only by the difficulty of the task. The range of sub-central bodies had reached substantial proportions, far in excess of the government's capacity to manage them. Bowen[28] underestimates the cumulative increase in non-departmental bodies alone at 64 per cent between 1949 and 1975 and the Pliatzky Report,[29] which excluded the NHS and the nationalised industries, estimated that there were 2117 such bodies in 1979. The centre regularly experienced difficulty in imposing its policy preferences on these non-elected species of SCG.[30]

Other types of SCG were also 'incorporated' and, in the case of the nationalised industries, they were explicitly part of the 'Social Contract'. Once again their prices were controlled and, predictably, concern grew over the scale of losses. Ostensibly, the industries had commercial freedom but this freedom continued to finish a poor second to government views on the needs of the economy. Incorporation was a different label for the continuing failure to evolve a coherent pattern of ministerial intervention and control.

The threat from nationalist politics produced an almost incoherent response from the centre. The elections of 1974 made the Labour government dependent upon third parties (including the SNP) for its stay in office. These elections provided clear evidence of 'partisan de-alignment'[31] as the proportion of votes cast for third parties rose, not only in the periphery, but also in England. The Labour government required the support of the Liberals and the SNP, both of which supported devolution, to preserve a working majority.

This daunting problem was compounded by the parlous state of the British economy. The end of economic growth denied the government the compromises, so typical of functional politics, of marginal improvements in service provision. Both inflation and unemployment rates rose, and cuts in public expenditure were common. Between 1976 and 1979, the Labour government introduced a species of monetarism, the Conservative opposition was taking a decisive step towards social market liberalism and the belief in salvation through institutional modernisation was collapsing on its transparent failures. Whether these shifts are seen as a loss of confidence in government or the loss of confidence by governing elites, the period was one of reassessment and search for new directions. Moreover, the government was now denied the usual escape route of adventures in 'High

Politics'. Britain's role in the world had contracted immeasurably and the country was vulnerable to externally generated instability. Attention upon domestic economic ills was not easily distracted and the Labour government did not have its 'Falklands' to rescue its public esteem. Devolution was, therefore, a policy bred of central weakness rather than nationalist strength. Certainly the policy was much disliked by both ministers and civil servants. The legislation bears all the hallmarks of antagonistic reluctance, being both internally inconsistent and ambiguous (often at the same time).

The first point to note is that the problem of nationalist politics was factorised. Separate policies were pursued in Scotland, Wales and Northern Ireland. Moreover, there was a policy (of sorts) for England. 'Organic change', or the redistribution of functions between the different types of local authority three years after a major reorganisation, was first mooted in a devolution White Paper and preoccupied the Secretary of State for the Environment between 1977 and 1979. There was still no grand design for the reform of territorial politics, just a series of *ad hoc* responses.

Second, at no stage was the doctrine of the supremacy of Parliament under challenge. Throughout, the centre insisted, for example, on the retention of all powers of economic management. However, it is probably a mistake to suggest that constitutional issues were at stake. Political survival was the key consideration for a fragile government; the centre was to concede as little as possible commensurate with the Labour Party retaining power.

Third, although the growth of central intervention took a very poor second place to the devolution debate at the time, the changes in the form of that intervention were particularly significant. The definition of the centre's responsibility for economic management had stressed aggregate control. For local government, this responsibility encompassed the level of central grant to local government. From 1975–76, this responsibility was unilaterally redefined. Local *expenditure* (not grant) was explicitly included in PES, the anticipated level of expenditure for individual local services was identified and general guidance on the level of rate increases was proffered. In other words, central government was intervening to regulate local expenditure. It had not been so targeted before. Hereafter, the control was to become more specific.

Finally, as ever, there was the sorry plight of Northern Ireland. Direct rule via the Northern Ireland Office prevailed. Public expenditure was increased to offset economic decay. But the most striking facts are the continued insulation of Northern Ireland from mainland politics and the growing indifference of Westminster to maintaining the Union. Governing without a consensus and the failure of the several attempts to find a substitute for direct rule have bred 'contingent commitment'.[32] Whatever else Northern Ireland may illustrate, it highlights the *ad hoc* and variegated strategies of the centre in managing SCG.

Ostensibly a period of dramatic transformations, the years 1974–79 reaffirmed the resilience of the Union. If anything, nationalist politics were weaker in 1979 than at any time in the past decade. But there had been important changes. If the new forms of central interventions in IGR were not newsworthy, they had the potential for transforming the system. That potential was soon to be realised.

4. 1979–83: Economic Recession, Control and Centralisation

If one single theme permeates this period, it is the search by central government for more effective instruments of control (not influence) over the expenditures of SCG. Control is not, however, the only development of significance. The government has also sought to restrict the size of the public sector by privatisation. Although its 'progress' under this head was initially slow, none the less it became a distinctive feature of Conservative policy. Whereas control and privatisation can be seen as an attack on the public sector, as a means of curtailing its role in British society, the third feature of central policy in this period involved the proliferation of non-departmental public bodies, thereby expanding the public sector and making it more complex. Certainly, local government railed vigorously against its containment and its bypassing in favour of non-departmental public bodies. The consultation so characteristic of the mid-1970s was replaced by bureaucratic direction and confrontation.

For local authorities, government policy was to make local *income* and expenditure conform with *national* decisions. The achievements of the new system can be described, at best, as mixed. Between 1979/80 and 1982/83 under the Conservatives total expenditure rose as a proportion of GDP and central government's expenditure increased sharply. The big spenders were not for the most part local government services and some local services declined dramatically: e.g. housing. The Conservatives cut substantially the centre's contribution to local services – the proportion of net current local expenditure financed by block grant fell from 61 per cent to 56 per cent while the proportion of expenditure financed by specific grants rose. Local government manpower fell by 4 per cent in the same period. Most dramatically, local capital expenditure was subject to stringent regulation and was reduced by some 40 per cent in real terms from an already severely reduced base. And yet, in spite of the 'cuts', local current expenditure *increased* between March 1979 and March 1983 by 9 per cent in real terms. Even attempts to 'cull the herd' of non-departmental bodies were conspicuously ineffective, involving some 240 bodies and a mere £11.6 million. The savings actually made were more than offset by the expansion of other non-departmental bodies such as the Manpower Services Commission. Equally, the attempts to impose strict External Financing Limits (EFLs), or cash limited ceilings on the grants and borrowing from government, of the nationalised industries failed. Out-turn in 1982/3 exceeded the plans of 1980 by some £1.75 billion (in real terms). Only privatisation, or the special sales of public assets, had any success, meeting the targets set since 1982/83 and totalling some £1.2 billion in 1983/84.[33]

In short, as with the previous Labour government, the bulk of the cuts fell on local government, on selected services and on capital expenditure while, ironically, the Conservatives presided over both an increase in total public expenditure and local current expenditure.

The consequences of Conservative policy can be summarised under seven main headings: control, privatisation, unilaterism, litigation, uncertainty and risk avoidance, unintended consequences and recalcitrances. Each outcome will be considered *seriatim*.

Control: To employ Stewart's distinction[34] between intervention and control, government actions impinged on individual local authorities to variable effects, only some of which were related to the centre's intention of cutting local expenditure. Thus, the grant system took *seven* different forms between 1979 and 1983[35] and the government became locked in a pattern of repetitive legislation: that is, legislating for the unintended consequences of earlier legislation.

Privatisation: Whether in the guise of contracting-out, liberalisation, charging for services or the sale of public assets, the amount of privatisation was limited. The only important exception was the rise in council house rents and the sale of council houses. Instead of freeing resources for the private sector, this strand of Conservative policy served primarily to exacerbate relationships between the two levels of government.

Unilateralism: It is difficult to avoid the conclusion that prior agreement on policy was an essential prerequisite of consultation. Corporatist-style arrangements were rejected, the rules of the game were unilaterally abrogated and the logic of negotiation collapsed. The new directive style served to promote recalcitrance.

Litigation: With ever-increasing conflict, the two levels of government resorted to the courts to regulate their relationship. There has been an unprecedented number of key cases, and the incidence of litigious behaviour has prompted the search for 'judge-proof' legislation by the centre and 'legalism' at both levels of government as they sought to check the validity of their own and the other party's actions.[36]

Uncertainty and Risk-Avoidance: The combined effects of an ever-changing grant system, unilateral changes in the rules of the game and legal challenges created a climate of uncertainty in which budgeting becomes characterised by mid-year revisions, the use of large contingency funds and the continuous allocation (and re-allocation) of funds from it. In effect, local government did not know what monies it had to spend and the centre did not know what local authorities would spend. Uncertainty has also bred an attitude of 'never do now what you can leave until later' and led to the development of a variety of strategies for avoiding the consequences of ever-changing government policy: 'creative accountancy'.

Unintended Consequences: The continuous changes in the grant system and the failure to cut local current expenditure may be the most serious unintended consequences from the government's standpoint but they are not the only ones. Two examples will suffice. First, capital expenditure did not just fall sharply in line with government plans. The tight revenue squeeze on local authorities made them reluctant to incur capital expenditure generally and to use the revenue from capital sales in particular. Moreover, the receipts from such sales were larger than planned. As a result, capital under-spending fell below the intended levels, prompting (unsuccessful) calls from the centre for increased expenditure! Second, the resource squeeze on local authorities increased the cost of provision in other parts of the public sector. The mentally and physically handicapped remained in the more expensive national health service and were not transferred, in accord with government policy, to local authority care because of the resource squeeze on the personal social services. Similarly, the rent increases required for council houses fell, in many cases, on the Department of Health and Social Security.

Recalcitrance: As the centre abandoned consultation for unilateralism, so local authorities abandoned co-operation for confrontation. Thus, Labour-controlled authorities increased their rates to compensate for loss of grant and their expenditure regularly exceeded government targets. The full-blooded adversary response focused on subsidised public transport. Following the House of Lords decision on the Greater London Council's (GLC) 'fare's fair' policy, the government introduced the Transport Act, 1982, to tighten its controls over public transport, especially the level of subsidy. This move was obviously partisan, provoking an antagonistic response and, once again, resort to the courts. Yet the level of subsidies remained high, an outcome which might seem surprising but for the fact it epitomises the fate of most Conservative policies on local government!

With an election imminent, the Conservatives faced a choice. They could intensify direction in spite of its manifold attendant unintended consequences or search for a more conciliatory mix of strategies. The (then) Secretary of State for the Environment, Tom King, seemed to be extending an olive branch by talking of abolishing central spending targets. The Department of Environment (DoE) civil servants cautioned against detailed control of local finance and of dismantling the GLC and the metropolitan county councils (MCCs).

The Prime Minister wanted to 'do something' about local government, however.[37] No matter that the cabinet committee (MISC 79) under William Whitelaw's chairmanship had concluded that there was no viable alternative to the rates. Somewhat tentatively, it had suggested abolition of the 'overspending' GLC and the MCCs but Tom King was known to doubt the wisdom of such a move. Mrs Thatcher was dissatisfied and established a sub-committee of the Economic Committee of the Cabinet which endorsed the abolition proposals and, once again, reviewed the rates. But now there was an election and a gap in the manifesto waiting to be filled with the local government proposals. Enter the Treasury in the guise of the Chief Secretary, Leon Brittan, with a proposal to limit rate increases. Again, no matter that the proposal had been rejected by Cabinet: there was a gap to be filled and Mrs Thatcher, disregarding all known opposition within the government and the party, inserted 'rate-capping' in the manifesto. TINA ('There is no alternative') applied to local government as well as economic policy: direction remained the chosen path. The Conservatives now had an ill-conceived policy and, on winning the election, no Secretary of State. The previous incumbents were unrepentant opponents and there was no rush of volunteers for the job: enter Patrick Jenkin and 1984, disguised as a comedy of errors.

Strategies towards Scotland, Wales and Northern Ireland in this period are variations on the same theme. Scotland can be seen as the 'test-bed' for English legislation. Most of the grant innovations were introduced in Scotland first and, since 1981, the Secretary of State for Scotland has been able to reduce the grant to the individual local authority if he thinks its expenditure is 'excessive': a power subsequently available to the Secretary of State for the Environment. In Northern Ireland, grant to local authorities actually rose as the centre pursued a strategy of 'more of the same': i.e. proposing (unacceptable) alternatives to direct rule, using public expenditure to fight economic decay and insulating mainland politics from the problem. In Wales, a Welsh Consultative Council on Local Government Finance was introduced to implement the new grant system. Revealingly this innovation had initially been proposed as part of the Labour government's devolution package. It now reappeared as a means for control, not devolution. Both cases provide a wonderfully clear example of the centre factorising problems and using institutional reforms to solve its own problems.

5. To 1984 and Beyond

The story of the legislation on rate-capping and 'streamlining the cities' could alternatively be described as a 'black comedy' but such entertainment must be abjured for the more prosaic task of describing the outcome of the legislative process and, in a more speculative vein, the consequences of that legislation.

The central objective of the Rates Act 1984 is to take away the power of local authorities to determine their own rate level. For this purpose, local authorities fall into two groups: those subject to the selective limitation scheme and the rest, subject to a general limitation scheme to be activated, should the need arise, by the Secretary of State. To date, the latter has not been activated and will not be considered further.

Under the selective limitation scheme, the Secretary of State chooses the local authorities to be rate-capped. He must specify the criteria for his selection: that is, they must be 'high spending' authorities. The local authority must exceed government determined spending limits (known as GREA) or £10 million for the year and its expenditure must be 'excessive'. For 1985–86, 'excessive' meant that the authority exceeded GREA by 20 per cent and its target by a least 4 per cent. The Secretary of State then stipulates a notional level of total expenditure for each selected authority. The authority can appeal against this total; that is, seek a derogation. When the total grant available to local authorities is known, the expenditure limit can be translated into a maximum rate, taking into account the financial reserves of the local authority. If the two sides agree on this rate, it becomes the legal limit. Assuming that the local authority does not agree with this rate level, and that it has not been able to renegotiate it with the Secretary of State, the latter can lay an order before the House of Commons or impose, without parliamentary approval, an interim maximum rate. The local authority may not levy a rate which exceeds the maximum stipulated by the Secretary of State.

It is early days for describing the consequences of rate-capping. The major lesson of recent years is that the government does not always get what it wants. Thus, the first bout of rate-capping led to the government accepting the 'overspend' by the rate-capped local authorities for the past five years and, given that the total grant to local government was further reduced, they increased their share of grant at the expense of other, primarily county, councils. Nor were the reductions in expenditure particularly draconian. On the face of it, the rate-capped authorities had their existing (cash) expenditure frozen, although the GLC, ILEA and Greenwich faced a 1 ½ per cent cash reduction. However, individual authorities faced far more severe cuts, depending on the extent to which they had practised 'creative accountancy' in the past. No firm conclusions on the effects of the new regime on local expenditure will be possible until it has operated for a number of years. However, the government now has the powers to control both the income and expenditure of local authorities. Consequently, the conclusion that the post-1984 changes have sharply increased the degree of centralisation in British government seems as obvious as it is inescapable. There are three important caveats to this bald assessment.

First, the politicisation of central–local relations has increased apace. Historically, national and local political élites have been insulated from each other. In sharp contrast to France, for example, Britain has no tradition of politicians 'collecting' offices at the national and local level: of holding at the same time a seat in Parliament and a chairmanship of a local council. The budgetary crisis in Liverpool heralded a new era. The Secretary of State was drawn into negotiations with the city over its 1983/84 budget and, if in fact the minister made few concessions, he indisputably lost the political battle by appearing to make them.[38] Rate-capping has the potential to suck the Minister and the DoE into numerous one-to-one negotiating positions with local authorities willing and able to impose high political costs. Admittedly, this strategy was rejected by the 18 councils rate-capped in 1984/5 which preferred to act as a group and refused to seek a derogation. There are also risks for the local authority seeking a derogation: for example, the Secretary of State must be given all the information he wants in order to reach his decision and he can further reduce the permitted rate level. But the united opposition of the Labour councils crumbled as the deadline for making a rate became imminent and, for future rounds, attempts to bargain with the centre may appear more attractive. Such bargaining holds perils for the centre. To become embroiled in the budgetary details of 18 or more councils will strain its organisational resources; will provide opportunities for opponents to publicise errors and inconsistencies; and will lead to the Secretary of State imposing requirements to cut services and make redundancies – actions scarcely likely to enhance his popularity or gain him a good press. Obviously these remarks are speculative but the politicisation of central-local relations may well generate the most severe unintended consequences of the centre to date.

Second, the technical defects of the grant system have become increasingly apparent and, more important, the system has been attacked from unexpected yet authoritative quarters. Thus, the Audit Commission (1984) argued that the system is contributing to inefficiency and ineffectiveness in local government.[39] It criticised 'unnecessary uncertainties', the accumulation of reserves to counter uncertainties, 'serious distortions' in the allocation of grant, the complexity of the system and the lack of incentives to improve efficiency and effectiveness. It argued that a potentially sound system based on GREAs was 'being eroded because of, *inter alia*, the distorting effects of penalties and targets'.

Third, recent developments continue to generate instability and ambiguity.

Attacks on the grant system have been increasingly vociferous, and, on top of this system, the government will now impose a plethora of elected and indirectly elected bodies, in place of the GLC and MCCs. To describe the new system as complex is to resort to English understatement:

> it seems inconceivable to us that the new arrangements will produce a system which is more comprehensible and accessible to individual citizens. A two-tiered local government certainly has its problems but a system consisting of a host of separate joint boards and joint committees of dubious accountability, backed up by a new range of central government controls, all superimposed upon the existing district tier seems to us considerably more confused and problematical.[40]

It is possible to add to the list of consequences which will stem from this latest bout of legislation. For example, litigious behaviour has been given a further boost by the 1984 Act and 'legalism' has replaced consultation as the means for regulating central-local relations. But politicisation and complexity are the paramount unintended consequences.

Not only did the Conservative government intensify its centralising strategy, but it also sought to regulate non-departmental public bodies and to privatise public corporations. Thus, the government bypassed recalcitrant local authorities and created or expanded single-function non-departmental agencies. The Manpower Services Commission's role in providing training for employment mushroomed. The response to the crisis of the inner cities was a profusion of special agencies. The abolition of the GLC and MCCs produced a battery of QUELGOs or quasi-elected local government organisations. Hand-in-hand with this expansion went a series of controls. The government sought to regulate the finance and staffing of the agencies; to develop corporate management techniques; to contract-out specific services as a way of imposing financial discipline; and to liberalise services (e.g. electricity production, express coaches).

The most distinctive policy was privatisation. Although some initiatives were taken during their first term of office, it emerged to considerable prominence after 1983. A response to escalating public expenditure and frustration with the record of the nationalised industries, this policy was borne of opportunism rather than principled policy development. It has raised massive sums of money for the exchequer but, paradoxically, it has also fostered a *re*-regulation (not *de*regulation) of public industry. The government remains the largest shareholder with an effective veto over policy. It has also established new regulatory agencies – e.g. the Office of Telecommunications (OFTEL) – with more extensive (and precise) powers than existed before. In effect, therefore, government policy raises as many problems as it purports to solve.[41]

It would be foolhardy to deny that the thrust of the post-1984 reforms is centralising but, equally, it would be short-sighted to over-estimate the capacity of the centre to realise its objectives. Conservative policies contain the seeds of their own ineffectiveness. If centralisation does ensue, it seems certain that the system will be complex, ambiguous, unstable and confused. The resultant policy mess will lead inexorably to a further reorganisation of local government, if only for the centre to abolish a system in which their interventions fail to have the stated effect.

EXPLAINING THE CHANGES

To this point, the narrative has merely implied explanations for the changes in SCG. It is now necessary to be explicit. Figure 2 provides a summary explanation for the period 1945–74. Assuming a stable external support system, economic growth and a central élite ideology based on the mixed economy welfare state, the consequent expansion of the welfare state led to increased functional differentiation and the institutionalisation of the professions. The resulting policy networks lie at the heart of a system of functional politics which marginalised local political élites and supported the dual polity. This duality was further supported by both an external support system which still afforded central élites a role in High Politics and a two party system based on functional economic interests which subsumed other social cleavages to class and which failed to colonise the localities. In sum, functional politics came to dominate territorial politics.

FIGURE 2 HOW SUB-CENTRAL GOVERNMENT HAS CHANGED 1945–74

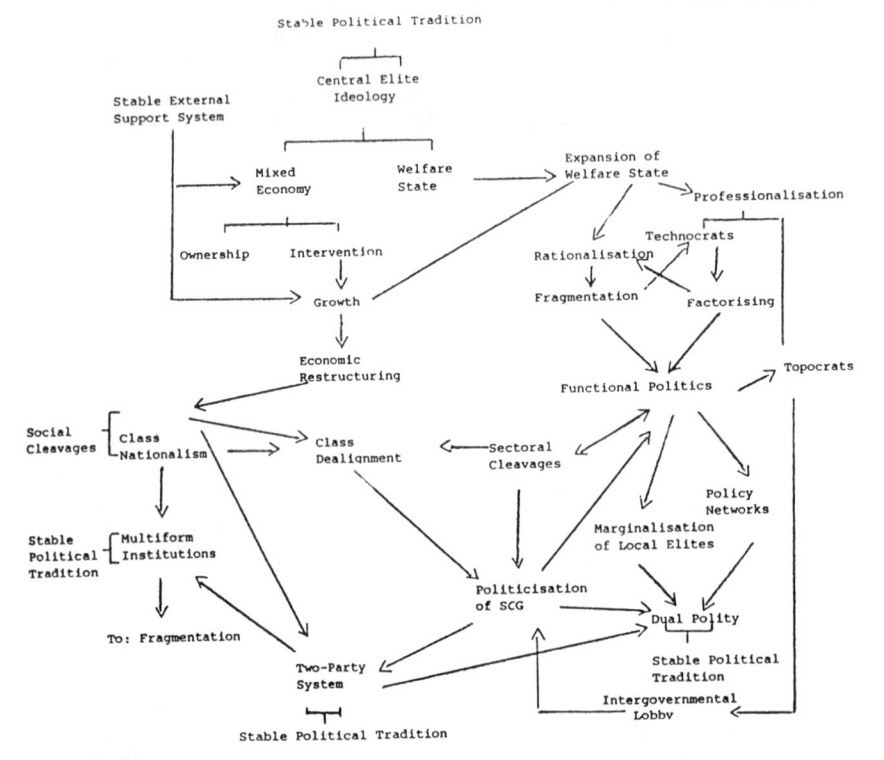

Arrows indicate causal relationships
Lines indicate sub-divisions of concept

However, there were stresses and strains within the system of functional politics. Economic growth and subsequent occupational restructuring fostered class dealignment. Traditional social cleavages re-emerged (e.g. nationalism) and functional politics generated sectoral cleavages. Moreover, Union had established a unitary state with multiform institutions and the expansion of the welfare state had further fragmented service delivery systems. This complex of organisations became the locus within which the conflicts of a social structure characterised by multiple cleavages were played out. Economic growth and functional politics did not herald the homogeneous society but increased the degree of social and political differentiation. Functional politics fostered a territorial reaction.

It is important to remember that the politicisation of local government began in an era of relative stability and prosperity. With deepening economic decline, an increasingly unstable external support system and a shift in central élite ideology, there was a quantum leap in politicisation. In addition, the problem-solving capacity of the system began to experience ever-intensifying difficulties. The planning experiments, whether in the guise of the public expenditure survey (PES) or of local corporate planning, were substantially abandoned and the compromises fostered by economic growth and incremental service expansion were now denied to the centre. If the centre's (in)ability to solve problems remained unchanged, the opportunities to avoid solutions by moving money were shrinking rapidly.

The changes between 1974 and 1985 can also be summarised in an explicitly explanatory form. Figure 3 identifies the major causes of changes in SCG under the exigencies of economic decline.

With the increasingly apparent failure of successive governments to manage the mixed economy successfully, the post-war consensus on Keynesian demand management, intervention, ownership and the welfare state began to founder. In its place, the 'New Right', concerned with the money supply, the PSBR and the minimal state, emphasised reductions in public expenditure and the contraction of the welfare state. This 'resources squeeze' intensified competition between policy networks determined to preserve their turf. Clients were mobilised and direct contact with local political elites was established. The 'hands-off' character of the Dual Polity was replaced by a directive territorial operating code. Nowhere was this change more obvious than in the changes to the system for distributing grant to local authorities but it is also evident in the efforts to reduce the number of quangos and in the investment and pricing controls over the nationalised industries. But direction was a high cost strategy. At first, the centre had channelled its political contacts through an intermediate tier of representation based on the national community of local government. Increasingly, this channel was replaced by face-to-face contact between ministers and individual (or small groups of) local authorities. Directive strategies provoked recalcitrance and non-compliance and, reluctantly, ministers were dragged into conflict with the New Urban Left and 'overspending' local authorities: conflict which spilt over, ever more commonly, into litigation. The costs in time and even loss of political face were mounting. To make matters worse, the policy was failing. The planned cuts in total public expenditure did not materialise. The new central elite ideology had not

FIGURE 3   ECONOMIC DECLINE AND SCG

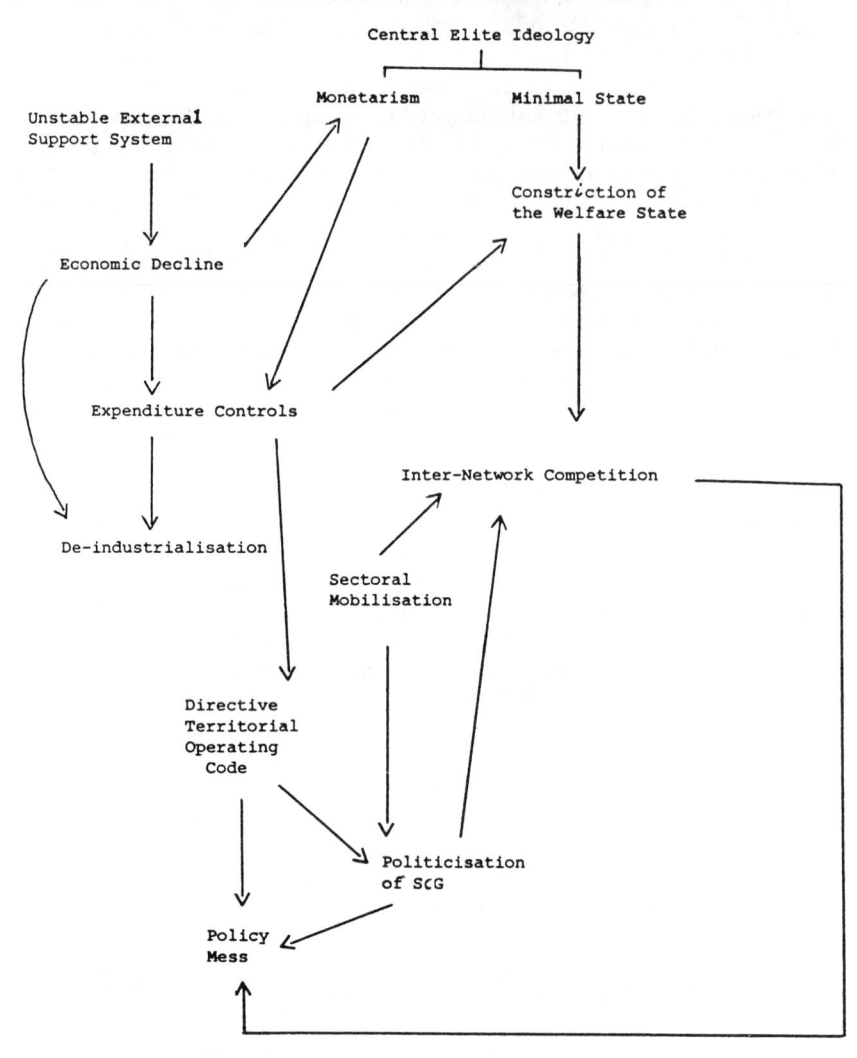

```
                        Central Elite Ideology
                        ┌──────┴──────┐
                    Monetarism      Minimal State
Unstable External                        │
Support System                           ▼
      │                           Construction of
      ▼                           the Welfare State
Economic Decline
      │
      ▼
Expenditure Controls
      │
      ▼
De-industrialisation      Inter-Network Competition
                              ↗       ↑
                         Sectoral
                         Mobilisation
                              │
                              ▼
                        Directive
                        Territorial
                        Operating
                        Code
                              │
                              ▼
                        Policy      Politicisation
                        Mess        of SCG
```

Arrows indicate causal relationships
Lines indicate sub-divisions of concept

countered economic decline with a reduction in the size of the public sector but it had intensified the politicisation of SCG and generated a policy mess wherein nobody achieved their objectives.

Moreover, the changes in SCG began to exert an increasing influence on the national government environment. These feedback effects are summarised in Figure 4. Thus, expenditure controls fell largely on local capital expenditure and, in effect, off-loaded public sector cuts on to the private sector — e.g. the construction industry. This simply contributed to the de-industrialisation

FIGURE 4 THE IMPACT OF SCG ON THE NATIONAL GOVERNMENT ENVIRONMENT

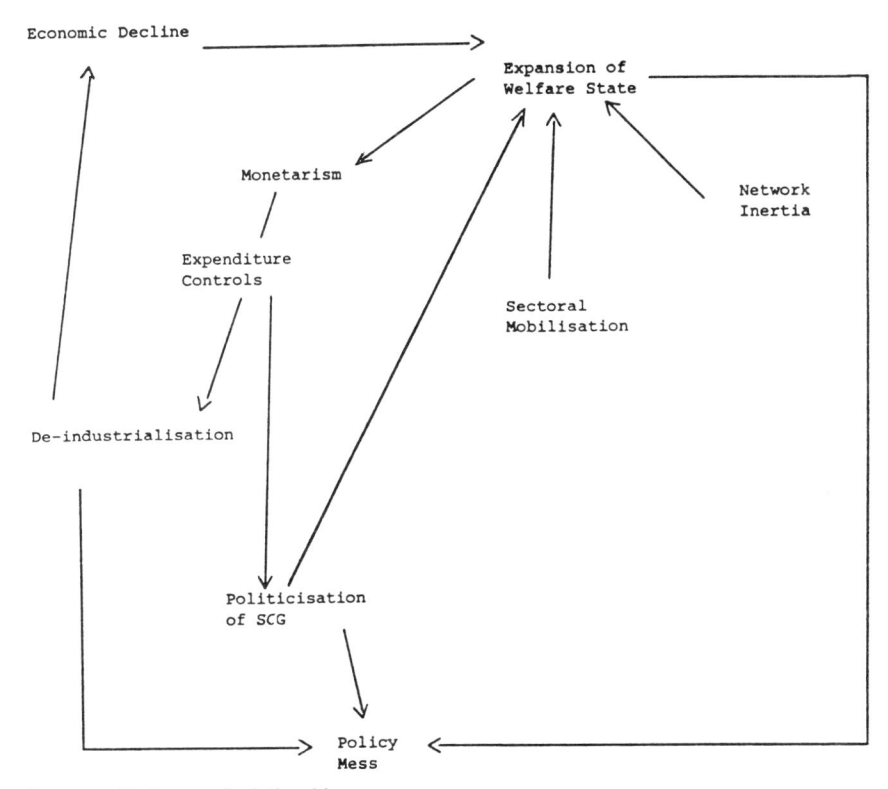

[Figure: Causal diagram showing relationships between Economic Decline, Monetarism, Expenditure Controls, De-industrialisation, Expansion of Welfare State, Network Inertia, Sectoral Mobilisation, Politicisation of SCG, and Policy Mess]

Arrows indicate causal relationships

of the British economy. Government action was intensifying economic decline which, in turn, increased the pressure on the welfare state, most notably social security payments to the unemployed. Coupled with the difficulties of making cuts because of both the capacity of some integrated policy networks to resist the pressure for reduced expenditure and increasingly strident protest from client groups, welfare state expenditures continued to rise. This increase threatened the government's monetary targets; intensified the search for means to control public expenditure; and generated yet more recalcitrance in SCG — thus the downward spiral continued. The resultant policy mess casts doubt on the efficacy of the ideology of central élites and thereby further contributes to partisan conflicts and the politicisation of SCG.

This summary explanation of changes in SCG is bald in the extreme. In reality, the sequences of cause and effect are rarely so clear-cut. But a simple, even simplistic, presentation provides both a sharp cutting edge for opening up the patient and, in the course of a further operation, the complications can be noted and the initial diagnosis can become both more thorough and exact. Many of the necessary caveats and modifications to the preceeding model have to be omitted.[42] More positively, there is a major advantage to a skeletal presentation: it is by its nature argumentative, challenging accepted accounts of SCG. The final section draws out the implications of the analysis for understanding territorial politics in the UK.

CONCLUSIONS: THE DIFFERENTIATED POLITY

This article has attempted to demonstrate that differentiation, disaggregation and interdependence – characteristics subsumed under the label 'the differentiated polity' – are of equivalent importance to parliamentary sovereignty, cabinet government and prime ministerial power (characteristics conventionally attributed to 'the unitary state') for the analysis of British government in general and its territorial politics in particular. By way of conclusion I propose to contrast as sharply as possible the images of the unitary state and the differentiated polity. The dichotomies listed in Figure 5 serve not only as a summary of much of the foregoing but, because they encapsulate the defects of the unitary state model, they also identify the continuing contradictions of British government. The figure provides a critique of the conventional portrait of British government and an explanation of the policy decline of the last decade; of the centre's faulty map of British government and the policy mess that is SCG.

The first point illustrated by Figure 5 is the range of contradictions in the differentiated polity. Changes in SCG in the post-war period were not caused by any one process – e.g. economic decline – or any simple combination of processes – e.g. economic decline and the resurgence of ideology. To understand the twists and turns of policy, especially over the past decade, it is necessary to explore the tensions between authority and interdependence, bureaucracy and differentiation (and disaggregation) and territory and function. And even this summary is an over-simplification for each of these tensions can be decomposed into more specific contradictions.

FIGURE 5   THE CONTRADICTIONS OF THE CENTRELESS SOCIETY

Unitary State	Differentiated Polity
1. *Authority*	*Interdependence*
Ideology	Domain consensus
Partisanship	Logic of negotiation
Accountability	Ambiguity
2. *Bureaucracy*	*Differentiation*
Co-ordination	Fragmentation
Command	Disaggregation
Direction	Inertia
Control	Intervention
3. *Territory (Dual Polity)*	*Function*
Factorising	Nationalisation
Autonomy	Professionalisation
Localism	Institutionalisation
Insulation	Politicisation

SCG operates in a system characterised by a strong executive tradition and can be no more divorced from the effects of the larger system than any other facet of British politics. The relationship between the centre and SCG is one of 'asymmetric' interdependence but, equally, the combination of differentiated service delivery systems and the hands-off tradition of central departments has fostered complex patterns of interdependence. Central to any understanding of central-SCG relations is the recurrent tension between interdependence of centre and locality on the one hand and authoritative decision-making by central government on the other. If the centre determines unilaterally the parameters of local action, then SCG can delay and frustrate the impact of central intervention. At the end of this road lies abolition but the journey is tortuous, exacts high political and administrative costs and, for all its much vaunted determination, the Conservative government was no nearer to resolving the tension between authority and interdependence than any of its predecessors.

Partisanship and ideology are crucial to understanding the *form* of the centre's response to the contradiction between authority and interdependence. The two-party system processes issues on a national stage and in an adversarial style which disrupts the logic of negotiation. It encourages the imposition of policies and strategies rather than permitting co-operation to emerge from the interplay of network interests and a sense of mutual advantage. The sheer visibility of issues encourages network participants to 'stand up and be counted', working against circumlocution and indirection, the avoidance of 'difficult' problems and the focus on means (rather than ends) necessary for compromises to emerge. The resurgence of party ideology has accentuated these effects. If both major parties have sought to control local expenditure, only the Conservative party, with its social-market liberalism, has sought to privatise parts of the public sector. Ideology has been the grit in the well oiled machinery of the policy networks which has disrupted domain consensus and challenged not only the assumption of inevitable incremental growth but also existing values and associated policies.

The result of these contradictions has been to undermine further the effectiveness of the traditional mechanisms of accountability. Their focus on specific institutions and their processes does not 'fit' the messy reality of a policy-making process which spans interdependent organisations. In spite of protestations and (ostensibly) policies to the contrary, confusion and ambiguity have remained pre-eminent characteristics of IGR over the past decade. The doctrines of ministerial accountability to Parliament and of collective cabinet responsibility belong to a simpler era when the mace did not confront the maze of the interdependent, differentiated polity. Yet these constitutional myths continue to be acted out as successive governments confront massive policy slippage with no means for calling to account those actors delivering the services and as they are called to account for their 'promises'. Accountability is confounded by ambiguity, breeding frustration and further bouts of speculative if not ineffective action.

If the national political élite is bemused by those constraints on its capacity to realise its goals, the administrative élite concerned to give effect to these goals has to confront the contradiction between bureaucracy and differentiation.

The picture of the centre as a co-ordinated machine able to translate intentions into facts disintegrates in the face not only of disaggregated service delivery systems but also of fragmentation at the centre. Historically, the army and the Roman Catholic church have been the exemplars of organisation and management theory: the efficient line bureaucracies. Whether or not this characterisation is (or ever was) accurate, it has exercised a pervasive influence over conceptions of bureaucracy; to the extent that the analogy with 'tools' and 'machines' is commonplace, even 'natural'. It is a profoundly misleading view of bureaucracy in British government. As Vickers[43] has repeatedly emphasised, the bulk of activity in government is not 'goal seeking' but consists of regulating changes in our relations by setting and resetting norms (or standards). The bureaucratic model is a narrow appreciation of relations in SCG which has imposed a succession of specific goals and stuck rigidly to a set of control norms. The failure to reset norms, to appreciate SCG as a set of relations to be regulated over time, is another way of describing the centre's territorial operating code as faulty.

The paradox of British government is that the tradition of 'leaders know best' coexists with weakness at the centre: co-ordination *versus* fragmentation. This weakness 'is struggled against manfully by the Treasury, CSD, Cabinet Office, and CPRS. But it is a debility all the same'. This debility is government by a committee in which 'most committee men are also the chief executives of their own departmental empires'; 'Everyone knows they serve themselves by serving their departments.'[44] Underlying the twists and turns of central policy is the simple fact that central government has many interests. Intervention may serve the interests of the Treasury (and even the DoE) but not so other spending departments. The several chief executives and associated empires are each embedded in a network which is not only a constraint but also a support in the 'manful' struggles against central co-ordination. Policy co-ordination is confronted and confounded not simply by a determined minister but by the *structure* of the centre with its diversity of interests enshrined in discrete networks.

Just as co-ordination has to overcome central fragmentation, so central commands have to overcome disaggregation. The weakness at the centre is re-enacted in relations with SCG. A centre with 'hands-off' controls attempts to regulate a plethora of organisations which have 'hands-on' controls over services of key importance to the centre. The combination of 'hands-off' central controls with disaggregation has fostered complex patterns of interdependence. The roots of this contradiction lie in the inception of the welfare state. The use of local government as the vehicle for delivering the services coupled with the predilection for 'ad-hocracy' created a range of disaggregated service delivery systems. The bureaucratic notion of command, with its corollary of a line hierarchy of superiors and subordinates, is ineffective in a context where there is no clear 'line'. Disaggregation calls for structures of incentives and sanctions tailored to particular policies and the structures of their networks. In a disaggregated context, command becomes a euphemism for policy slippage.

Policy slippage is increased to the degree that the centre seeks to be the source of policy innovation and direct SCG. As the Treasury has found to its cost, some policy networks have fought a rearguard action of consequence against 'cuts'. It has had to confront the unwelcome fact of life that the alliance of professions and spending departments in policy networks has generated substantial inertia. Dynamic conservatism results in inertia and frustrated expectations. Innovation is possible – for example, by action at the boundaries of networks and by occupying new policy space – but not through the direction of existing networks without considerable costs in terms of delay.

The combination of central fragmentation, disaggregation and inertia leads to the contradiction between control and intervention. As Stewart[45] points out, the concept of control is underpinned by a model of the purposive organisation in which the centre has clear aims and control exists to the degree that these aims are achieved. But the fragmented centre has multiple goals and 'its' actions are better described as interventions which impinge on SCG but do not necessarily achieve central purposes. Moreover, viewing central action as intervention leads inevitably to a discussion of unintended consequences; to the state of affairs in which central action does not achieve its stated purposes, proliferates unintended consequences and thereby intensifies the problem of control which the initial central actions were designed to resolve. The resultant policy mess stands as testament to the inefficacy of a bureaucratic model of action in a differentiated polity: a consequence exacerbated by the demise of the dual polity.

For the bulk of the post-war period, the centre's territorial operating code sought to maximise central autonomy by insulating the national political élite from territorial politics. The problem of territorial management was less resolved than avoided. This dual polity was undermined initially by the tensions between territorial and functional politics and subsequently by the direct action of the national political élite.

To reinforce central autonomy and maintain the dual polity, the problem of SCG was factorised with separate policies for Scotland, Wales and Northern Ireland. Separate treatment can be interpreted as reinforcing national identities but, in the UK context, it facilitated the pre-eminence of Westminster and Whitehall. By accommodating nationalist demands, the dominance of the centre was perpetuated. There was no single problem of territorial protest but a series of discrete or factorised problems to be managed by institutional change and the redistribution of public expenditure. Conflict was diffused, a conclusion which applies equally to the modernisation of SCG in England. But if it had low visibility, the tension between territorial and functional policies remained ubiquitous. With the creation of the welfare state came not only differentiation but also the presumption that territorial justice required national or uniform stands.

The expectation of equality of access for all to services of equivalent standard throughout the country required the co-existence of fragmentation between networks and centralisation within them; of factorised treatment with national standards. The resulting tensions were manifested in the contradictions between autonomy and professionalisation and between localism and institutionalisation. Differentiation and professionalised policy systems grew hand in hand. The traditional conception of local authorities making and adopting policy for local needs and problems had a rival in the professional acting on the national stage and looking to shared notions of best professional practice for policy guidance and innovation. The professions are an early example of non-local sources of 'local' policy but they were followed by local politicians, especially after the reorganisation of local government. The burgeoning role of the national community of local government saw the institutionalisation of local politicians at the national level. Indeed the contradiction between localism, of local policies to meet local needs, and the generalising of local authority experience by its national representative bodies to meet the legislative needs of the centre has plagued the national community over the past decade.

The combination of nationalisation, professionalisation and institutionalisation served to undermine the mutual insulation of centre and locality which lay at the heart of the dual polity. The decline was accelerated by the actions of the centre. The onset of intensive intervention led to the adoption of a command operating code at variance with the differentiated polity. Not only did the centre lack the capacity to control SCG but its reliance on bureaucratic strategies multiplied unintended consequences and accelerated the politicisation of SCG. This faulty operating code may have its roots in the centre's 'culture of disdain'[46] but it is more plausibly a product of the dual polity in which the two levels of government were insulated from each other and of central fragmentation (in which the policy networks are insulated from each other). When structure imposes blinkers, cultural explanations are embellishment and detail. The command code represents a failure to comprehend that British government is differentiated and disaggregated: the unitary state is a multiform maze of interdependencies. To operate a code of variance with this reality is to build failure into the initial policy design.

The clearest indicator of the demise of the dual polity and the collapse of insulation is the politicisation of IGR. Since 1979, it has proceeded apace and the government's determination to control individual local authorities has brought ministers into face-to-face conflict with local leaders. Perhaps the most significant feature of the refusal of Liverpool City Council to make a legal budget was the *bargaining* between the Secretary of State and local leaders. Rate-capping is, potentially, a catalyst to such interaction between national and local élites. And this politicisation also demonstrates that IGR are no longer government and politics 'beyond Whitehall' but they are part and parcel of the politics of Whitehall. These changes are probably the most important ones in the British system of IGR during the 1980s and their full weight remains to be felt.

It is obvious that national economic management problems have dominated the practice and reform of intergovernmental relations. Local expenditure decisions were increasingly determined by central government's public expenditure survey system and the attendant Cabinet decisions throughout the 1970s. Many of the rapid changes in the expenditure targets of local authorities can be attributed directly to the decline of the British economy. Intergovernmental relations have become inextricably entwined with economic management whether the government of the day is Labour or Conservative. The analysis of SCG does not proceed very far, however, if the explanation of changes since 1945 remains confined to their economic context.

Any understanding of developments in SCG has to be rooted in an analysis of the range of sub-central organisations and the evolution of their relationship with the national government environment. From this perspective it is possible to explain the demise of the Dual Polity, why the Conservatives sought control and why their search experienced so many problems. Developments over the past decade suggest that control strategies are at variance with the Dual Polity and, when preferred, the outcome will be unintended consequences, recalcitrance, uncertainty, ambiguity and confusion. Britain may be in the early stages of developing a 'Napoleonic code'. A non-executant centre seeking 'hands on' controls will need to develop eventually an organisational infra-structure for exercising such control. The emergence of expanded regional offices of the DoE charged with vetting the budgets of rate-capped local authorities was an unthinkable idea ten years ago but, today, it seems the next logical step. And yet such a growth in the number of bureaucrats is an anathema to a political élite wedded to a populist electoral strategy. The very existence of such divergent, even contradictory, strands attests to the disarray in the central operating code. To a considerable degree, the government has been and remains a prisoner of the interests it seeks to regulate for those interests are part of the structure of government. And in seeking to escape, government actions have politicised SCG, further eroded the Dual Polity without finding a substitute territorial operating code and intensified the problem of control. The history of SCG is compounded of multiple contradictions – economic, political and organisational. Mono-causal explanations, whether propounded by politician or academic, are doomed to inadequacy. Policy-making for SCG has generated a policy mess because of the failure to appreciate that disaggregation, differentiation, interdependence and policy networks are central characteristics of the British polity which can be no more disregarded than the executive authority of the Prime Minister and the Cabinet or the role of Parliament.

NOTES

My thanks to Pat Dunleavy (LSE), Mike Goldsmith (Salford), Caroline Cunningham (Essex), Vincent Wright (Nuffield College) and Peter Saunders (Sussex) who offered helpful comments on an earlier draft. I should also like to acknowledge the financial assistance of the Economic and Social Science Research Council (Grant Number E002420014).

1. J. G. Bulpitt, *The Problem of 'The Northern Parts': Territorial Integration in Tudor and Stuart England* (University of Warwick: Department of Politics, Working Paper No. 6, 1975), p. 39; and by the same author, *Territory and Power in the United Kingdom* (Manchester: Manchester University Press, 1983), p. 134.
2. Inevitably in a summary paper of this kind I have drawn on earlier publications. The second section draws upon R. A. W. Rhodes, '"A squalid and politically corrupt process"?: Inter-governmental relations in the postwar period', *Local Government Studies* Vol. 11 No. 6 (1985), pp. 35–57 and Rhodes, *Beyond Westminster and Whitehall: the sub-central governments of Britain* (London: Allen & Unwin, 1988, forthcoming). The third section draws on Rhodes, 'Continuity and Change in British Central–Local Relations: the "Conservative Threat", 1979–83', *British Journal of Political Science* Vol. 14 (1984), pp. 311–33; Rhodes (1985); Rhodes, *The National World of Local Government* (London: Allen & Unwin, 1986); (with P. Dunleavy), 'Beyond Whitehall' in H. M. Drucker *et al.*, (eds.), *Developments in British*

Politics (London: Macmillan, 1983); (with P. Dunleavy), 'Government Beyond Whitehall' in H. M. Drucker *et al.*, (eds.), *Developments in British Politics 2* (London: Macmillan, 1986). The fourth section draws on Rhodes, '"A squalid and corrupt process ..."?' and *Beyond Westminster and Whitehall*. The Conclusions also draw on this latter book and 'La grande bretagne, pays du "gouvernement local"' *Pouvoirs* No. 37 (1986), pp. 59–70. My thanks to the various publishers and editors, and especially to Pat Dunleavy, for allowing me to draw on this material.

3. J. D. Stewart, *Local Government: the conditions of local choice* (London: Allen & Unwin, 1983), pp. 11–12 and 65–8.
4. J. K. Benson and Associates, 'A Framework for Policy Analysis' in D. Rogers and D. Witten (eds.) *Interorganizational Co-ordination* (Ames, Iowa: Iowa State University Press, 1982).
5. Bulpitt, *Territory and Power*, pp. 59, 137 and 167.
6. A. Singh, 'UK Industry and the World Economy: a case of de-industrialisation?' *Cambridge Journal of Economics* Vol. 1, (1977), pp. 113–36.
7. L. J. Sharpe, 'Is There a Fiscal Crisis in Western European Government? A First Appraisal' in L. J. Sharpe (ed.), *The Local Fiscal Crisis in Western Europe* (London: Sage, 1981), pp. 5–28.
8. N. Luhmann, *The Differentiation of Society* (New York: Columbia University Press, 1982), pp. xv and 353–5.
9. L. J. Sharpe, 'Decentralist Trends in Western Democracies: A First Appraisal' in Sharpe, (ed.), *Decentralist Trends in Western Democracies* (London: Sage, 1979), pp. 9–79.
10. The term 'technocratic profession' refers to programme of functional specialists. The topocratic professions – from *'topos'* meaning place and *'kratos'* meaning authority – cover the range of functions and speak for their areas. See: S. Beer, 'Federalism, Nationalism and Democracy in America', *American Political Science Review*, Vol. 72, (1978), pp. 9–21.
11. Bulpitt, *Territory and Power*, pp. 64–5 and 149–52.
12. I. Crewe, 'The Electorate: Partisan Dealignment Ten Years On', *West European Politics*, Vol. 6 (1983), pp. 183–215.
13. P. Dunleavy, *Urban Political Analysis* (London: Macmillan, 1980), pp. 42–55 and 70–96.
14. R. Rose, *Understanding the United Kingdom*, (London: Longman, 1982), p. 50.
15. R. Plant, 'The Resurgence of Ideology' in H. M. Drucker *et al.* (eds.), *Developments in British Politics* (London: Macmillan, 1983), pp. 7–29.
16. M. Kogan, *The Politics of Education* (Harmondsworth: Penguin Books, 1971), p. 171.
17. Bulpitt, *Territory and Power*, pp. 68 and 165–7 and 235–6.
18. P. Arthur, *Government and Politics of Northern Ireland* (London: Longman, second edition, 1984), p. 83.
19. J. A. Brand, *Local Government Reform in England* (London: Croom Helm, 1974); Dunleavy, 'Quasi-governmental sector professionalism' in A. Barker, *Quangos in Britain* (London: Macmillan, 1982).
20. E. Page, 'Grant Consolidation and the Development of Intergovernmental Relations in the United States and the United Kingdom', *Politics*, Vol. 1, No. 1 (1981), pp. 19–24.
21. For a more detailed discussion, see Bulpitt, *Territory and Power*, pp. 146–55.
22. J. G. Kellas, *The Scottish Political System* (London: Cambridge University Press, 1975), pp. 31–40.
23. Rhodes, *The National World of Local Government*, Chapter 6.
24. Rose, p. 88.
25. D. W. Urwin, 'Territorial Structures and Political Developments in the United Kingdom' in S. Rokkan and D. W. Urwin, (eds.), *The Politics of Territorial Identity* (London: Sage, 1982), p. 68.
26. Layfield Committee, Committee of Inquiry into Local Government Finance, Appendix 1, *Evidence by Government Departments* (London: HMSO, 1976), p. 327.
27. This section is drawn from Dunleavy and Rhodes, 'Beyond Whitehall', pp. 125–6.
28. G. Bowen, *Survey of Fringe Bodies* (London: CSD, 1978), p. 1.
29. Pliatzky Report, (London: HMSO 1980), p. 5.
30. See the essays in B. Hogwood and M. Keating (eds.), *Regional Government in England* (Oxford: Clarendon Press, 1982).
31. Crewe, 'The Electorate'.
32. Rose, p. 218.

33. P. Riddell, *The Thatcher Government* (Oxford: Martin Robertson, 1983).
34. Stewart, p. 61.
35. G. Jones and J. Stewart, *The Case For Local Government* (London: Allen & Unwin, 1983), p. 37.
36. M. Grant, *Rate Capping and the Law* (London: Association of Metropolitan Authorities, 1984).
37. A. Forrester, S. Lansley and R. Pauley, *Beyond Our Ken* (London: Fourth Estate, 1985), pp. 64–6.
38. A. Midwinter, 'Setting the Rate – Liverpool Style', *Local Government Studies*, Vol. 11, No. 3 (1985), pp. 25–33.
39. The Audit Commission, *The Impact on Local Authorities' Economy, Efficiency and Effectiveness of the Block Grant Distribution System* (London: Audit Commission, 1984).
40. N. Flynn and S. Leach, *Joint Boards and Joint Committees: an evaluation* (University of Birmingham, INLOGOV, 1984), p. 42.
41. For a more detailed discussion see Dunleavy and Rhodes, 'Government Beyond Whitehall' pp. 107–43.
42. Rhodes, *Beyond Westminster and Whitehall* ...
43. Sir Geoffrey Vickers, *The Art of Judgement* (London: Chapman and Hall, 1965), p. 33.
44. H. Heclo and A. Wildavsky, *The Private Government of Public Money* (London: Macmillan, 1974), pp. 369 and 371.
45. Stewart, pp. 60–62.
46. R. Greenwood, 'Pressure from Whitehall' in Rose and Page (eds.), *Fiscal Stress in the Cities* (London: Cambridge University Press, 1982), p. 72.

France: The Construction and Reconstruction of the Centre, 1945–86

Yves Mény

J. Gottmann has pointed out that, while the concept of territory is, in itself, a neutral notion, its political significance derives from the interpretation and values conferred upon it by a population (for ethnic, historical and linguistic reasons).[1] In the French case, history, economy and ideology have combined to give territory and its organisation such symbolic and concrete importance that France has been presented as the prototype of the centralisation and control of territory from a single centre through a vertical and hierarchical system. In essence, this impression remains valid even if it is based too often on the appearance and form of a system which, in reality, is considerably more complex. However, the centralisation and dominance of Paris (to a point where the French find it difficult to imagine a non-Parisian form of geographic centralisation or to conceive that the functions of a capital could be fulfilled by a medium-sized town) has neither the absolute, nor linear character which is often too easily attributed to it. This contribution examines the development of the relationship, the checks and balances, between centre and periphery in France during the period 1945–86.

Four phases can be identified between 1945 and 1986. They do not, however, fit with normal constitutional and political chronology (Fourth and Fifth Republic and the de Gaulle, Pompidou, Giscard and Mitterrand presidencies). The first (1945–54) was a period of aborted chances and immobility. The second (1954–69) was characterised by the intensity of both political and social change. The third (1970–1978) was marked by a shift in the strategy of central government and by the strong influence of local notables on the definition and implementation of policies. The most recent period (1978–86) has been one of economic recession during which local governments have been very active, and central governments keen to find an agreement with the local authorities in order to transform central–local relations. After coming to power in 1981, the Socialists succeeded where Giscard d'Estaing had failed, in achieving a decentralisation reform.

1945–54: THE DREAM OF THE GOOD OLD DAYS

The most striking feature of the immediate post-war period was the atmosphere of restoration of the old republican virtues, strangely associated with the willingness profoundly to reform the political, economic and social structures of the country. The political influence of the periphery was weakened by several factors: the Radical Party and the right were severely undermined because of their links with the Vichy regime and Marshal Pétain. Those political leaders who had not been banned from electoral competition were in any case considered responsible for the defeat and too closely associated with the Germans and their French collaborators. The situation was even worse for the nationalist parties which had been favoured by the Nazis and had been among the warmest supporters of the Vichy regime. Regionalism and regionalisation were equated with collaboration and fascism. It took years and a new generation of regional activists to dissociate the regional claims from a particular phase of recent French history.

The other political parties were either committed to the tradition and virtues of political and administrative centralisation (Socialists, Communists), or too weak at the local level to be the appropriate channel for peripheral demands. In the field of local government, the programmes of dominant political parties were quite limited and loosely characterised by the catch-word democratisation. Its meaning was two-fold: first, it expresssed the will to eliminate all elements linked to the Vichy regime and especially its limited and piecemeal attempts to regionalise. Instead, the French local system should return to its pure republican essence: a two-tier system of *départements* and *communes*. Second, it meant a reinforcing of the democratic element by transferring the Prefect's local powers to an elected executive, the President of the *Conseil Général*. But both this constitutional reform and the special status promised (but not delivered) to the big cities never got beyond the drawing board.

The 1947 cold war (and the exclusion from government of the Communists) and the pressure of colonial rebellions postponed any potential change. But in the meantime, the political coalition which governed the country up to 1947 had effectively reinforced the centralising trends of the past, notably through increased state intervention. Christian Democratic support for social justice and economic control coincided with the more marked preferences of the Communist Party and the Socialists for nationalisation. Finally, the central role of the state in Gaullist doctrine justified the Jacobin and Colbertist-type measures undertaken under the auspices of General de Gaulle: centralised economic planning, a national social security system, nationalisation of large banks, coal mines, gas, electricity, and transport companies. By the 1950s, France was more centralised than ever, both politically and economically, and no attempt at all (except for the symbolic and unsuccessful attempt to abolish the prefecture) was made to reshape central-local relationships. The title of a book published in 1947 summarised the situation: *Paris et le désert français*.[2]

1954–69: THE MAELSTROM OF REFORM

If France seemed to have been a kind of 'sleeping beauty' as far as its social structure and reforms were concerned, that was certainly not true of the economic or political spheres. Governmental crises were a permanent feature of the Fourth Republic for internal and/or external reasons, and economic growth challenged the traditional pattern of employment and population distribution. Up to the Second World War, France was predominantly rural (only half of the population lived in towns, i.e. *communities of more than 2000 inhabitants*!) and nearly one worker out of four was occupied within the agricultural sector. The exodus from the countryside to cities, which it had been possible to slow down during the war because of difficult urban living conditions and the ruralist policy of the Vichy regime, started up again after the Liberation. So, to the rebuilding of destroyed houses was added the task of housing thousands of migrants, not to mention all those living in substandard or unhealthy conditions. It was not until the 1970s that this immense effort began to eliminate the housing crisis. Nearly 10 million housing units and thousands of schools and hospitals were built during the thirty years following the war and a large part of the burden had to be shared by local government.

In the meantime, the discrepancies between the rapid growth of some regions or cities (mainly Paris) and the decline of entire rural regions (Brittany, Corsica, Massif Central, the South West) became more acute and visible. Economic planning, being mainly concerned with sectoral development and big projects, far from attenuating the natural trends, accentuated the phenomenon. This situation created the ideal terrain for protest. But it is remarkable that protest movements were not channelled by the local élites, the famous *notables* who were more interested in maintaining the status quo than in promoting economic or social change. The awakening of local and regional consciousness was the result of the mobilisation of the population in depressed areas by new élites such as union leaders, university professors, young professionals or businessmen. The reform era began, thanks to these various regional pressures and to the political impetus initially given by Prime Minister Pierre Mendès-France to reformism and change.

The Mendès-France government lasted only a few months. But this was sufficient not only to launch several reforms at the regional level, but also to create, among some civil servants of the central administration, a new state of mind: reform was becoming the watchword and, in spite of the weakness of successive governments, many changes of an incremental nature were introduced.

In this area, as in many others, the accession of General de Gaulle to power in 1958 and the creation of the Fifth Republic allowed the state to achieve many objectives which previously had failed through lack of political will. Within the first six months of the new Republic, the Prime Minister Michel Debré took (by means of executive decrees) a large number of decisions related to local government: encouragement of communal co-operation and regrouping, local financial reform, and incentives for creating metropolitan authorities. The spirit was decisively interventionist (not to say authoritarian), but the reforms were in fact rather limited, in order not to frighten the local *notables* that the Constitution had promoted to the rank of *grands électeurs* of the head of state. The failure of this first attempt made it necessary to take more drastic decisions. The Paris region was reorganised by law in 1961, and four large cities (Lyons, Lille, Marseilles, Strasbourg) were forced to enter into co-operation with their suburbs in 1966 (*communautés urbaines*). In the meantime, mergers were encouraged, although fewer than 2,000 communes (out of some 39,000) had been suppressed by the end of the 1960s, and the fiscal reform decided in 1958 was only finally enforced by 1967–68. The events of May 1968 had a double impact on central–local relations: they killed the Fouchet project, an ambitious plan to reduce the number of localities and force them to merge; and they justified − at least in de Gaulle's mind − the regional reform of 1969, proposed at the referendum, the failure of which put an end to his career.

From this 15-year period, in retrospect, five features emerge:

1. The intensity of reforms both in number and depth; thanks to political stability, the reforms were permanent and, in spite of some serious failures, the record of the period is impressive.
2. The deterioration of central−local relations: in order to avoid the resistance of local notables allied with the opposition (left or right), the Gaullists tried to build a new coalition based on the new local élites (the so-called *forces vives*). The failure of the 1969 referendum put an end to this risky strategy.
3. The 'heroic' style of reforms; emphasis was put on the unavoidable character of the reforms, and authoritarian measures were often preferred to more incremental policies.
4. In spite − or because − of the mutual antagonism between central government and local élites, a progressive nationalisation of the local system took place: local elections became national in character; the *cumul des mandats* system was strengthened and financial transfers from central budget to localities increased.
5. The most striking element concerns the multiple ways by which the central government succeeded in ensuring French economic and urban development during this period. Through grants, subsidies, incentives and prefectoral activism, the central government faced successfully the challenge of implementing national policies in spite of the continued existence of 37,000 communes.

1970−78: THE QUIESCENT PHASE

After ten years of wide-ranging reforms and the psychological and political shock of May 1968, the new President of the Republic, Georges Pompidou, felt the need for a pause in the process of change in order to build a new coalition which would ensure the support of local *notables*. This new course was the result both of political needs and personal convictions. Such a strategy was necessary, he thought, not only because the social basis of Gaullist support had become too narrow, but also because the industrial modernisation of France required a quieter management of other problems and the acquiescence of the political élites. However, there were also strong pressures for change from many political groups, be they radical (ultra-left) or reformist. In fact, Pompidou understood the ambiguous nature of these claims and put into practice the philosophy expressed by Lampedusa in his novel, *Il gattopardo*: 'Things must change to remain the same'. The two reforms launched in 1971 and 1972 in the field of local government were inspired by this strategy: to pay lip-service to the reformist ideal, while securing the maintenance of the status quo. In 1971, the Marcellin reform was geared, once again, to the elimination of small localities and to the reinforcement of co-operation between communes, especially in the metropolitan areas. At first sight, the law looked like a new and firm attempt to take a decisive step. In fact, the minister's instructions to the prefects and their 'complicity' with the local *notables* emptied the law of its substance. After a gigantic merger plan had been established (each *département* defined its own), the total number of communes was reduced by only 1,000.

The same spirit marked the regional reform. The 1972 law was conceived in such a way that the regions would remain, as Pompidou put it, *syndicats de départements* under the strict control of the prefect on one hand, and local *notables* on the other. The regional council was not directly elected and was composed of local members of Parliament and local officials, the prefect was in charge of the executive, and the regions were authorised (within a limited budget, its ceiling fixed by central government) to subsidise only state or local investments, being forbidden to have their own programme or policies. But once again, the implementation phase was decisive and ran contrary to the initial intentions of the anti-regionalist coalition. The new institution began to develop its own strategy, new leaders emerged and the central government appeared unable to control its own creature, especially where the opposition was in control of the regional council.

This policy of accommodation was pursued by Giscard d'Estaing partly for the same reasons: his main supporters were to be found among those who favoured the traditional institutional setting, and his followers — the *Républicains Indépendants* — are *notables* among the *notables*. To avoid conflicts with local officials, Giscard d'Estaing appointed a Commission chaired by Olivier Guichard (a former minister and local official) whose members were predominantly mayors or members of *département* councils (eight out of ten). However, the Guichard Commission still proposed rather radical solutions. Consequently, to appease most local officials who were upset and the left, which was hostile, Giscard d'Estaing failed to implement the Commission's proposals and simultaneously showed a greater reluctance to strengthen the regionalisation process. After the 1977 municipal elections, which strongly reinforced the left at the local level (two-thirds of the towns of more than 30,000 inhabitants were won by the left coalition), Giscard d'Estaing tried to find a way out by launching a huge consultation with the 36,394 mayors. On the basis of the 16,229 replies received, a report was prepared by a group of civil servants: a new period of reform was beginning.

In spite of this phase of relative immobility, some important innovations occurred: first, the institutionalisatin of regions which the central government was unable fully to keep under control; second, the increasing strength of the periphery, capable of extracting more and more resources from a central government in search of support; third, the substitution of incremental measures for the more 'heroic' style of the Gaullist period; fourth, the more and more frequent use of more flexible and less egalitarian tools for monitoring central policies (for instance, the development in many fields of the so-called 'contractual' relations); and lastly, a growing consensus on what the reform should not be and on the main components of a decentralisation policy.

1978-85: THE TRIUMPH OF THE PERIPHERY?

The 1977-78 consultation of the mayors was a major U-turn in the strategy of the central government towards the communes. For the first time, the central government did not try to impose its own views and aims on the periphery, but was willing to adopt reforms suggested by the local élites. After 20 years of unsuccessful attempts, the central government was giving up any kind of radical reorganisation, be it functional or territorial. The bill prepared by the Barre government was not discussed by the Senate until the eve of the 1981 presidential election. The text adopted by the Senate differed from the government's proposals and should then have been discussed by the National Assembly. Mitterrand's victory swept Giscard d'Estaing's project away, and the first reform bill introduced by his government was the *loi Defferre*, the so-called *grande affaire du septennat*. The *savoir-faire* of the new Minister of Interior was more evident in the strategy than in the content of the reform. Defferre had several cards in his hand: the opposition was extremely weak, and the left had a sweeping majority both at the national and local level. All conditions were met for imposing a radical reform.

Defferre did not try to introduce the kind of revolution announced by the *autogestion* programme of the Socialist Party. On the contrary, the Giscard d'Estaing bill amended by the Senate became the basis for the socialist reform. Most of the changes introduced by the socialist government were either accepted by the political élites or already introduced in practice. In some respects, the law was a simple codification of the practices of many left municipalities, *départements* and regions before 1981. The most striking changes – suppression of *tutelle* and the transformation of the prefects' functions – have only been partially implemented. Indeed, the law was conceived as a general framework applicable to every local authority. In fact, many of them are unable or unwilling fully to apply the new rules because they lack personnel or financial means, or they prefer the old prefectoral system. Apparently, the periphery has got what it wanted, and the most dynamic and powerful local authorities are able to take advantage of the changes introduced by the *loi Deferre*. For the others, however, business as usual remains the rule, and the central government still enjoys a vast potential for control and intervention.

It will be seen that, while France remains a centralised country, and the dominance of Paris is incontestable, the form and extent of central control have been modified under the pressure of the periphery. This evolution, which is not peculiar to France, contradicts those studies of the political development of modern industrial states that have followed a centralist and deterministic paradigm in which successful modernisation is equated with the integration of local institutions, social networks and political culture into a remarkably uniform social system.[3] Huntington[4] is a leading exponent of this approach to political modernisation, and his contention that 'the successful political modernization is the centralization of power necessary for social reform and the expansion of power necessary for assimilation' has been widely accepted. Thus, Lipset argued that 'the most dramatic form of resistance to modernising trends in post-industrial society has been the re-emergence of ethnic or linguistic nationalism in many countries'.[5] This interpretation of events seems valid only insofar as it refers to the development of Western societies in the 1950s or 1960s and is clearly based on subjective value judgements. On the contrary, at least in France, it seems that central government has been successful in adapting its structures and its relationship with the periphery. We will examine the ways by which centre and periphery have interacted and have been able to build up a new set of rules and institutions.

MINORITIES AND STATE: FROM MANY TO ONE?

Despite long standing historical roots, the problem of political identity remains one of the most important of the contemporary period. The notion of the nation-state which has prevailed and continues to prevail in France demands of the constituent parts of the nation an exclusive allegiance which tolerates no duality of political identity. The Jacobin state recognises neither infra-national nor supra-national identities. The modern French state has revealed itself to be as intolerant of national minorities as it is of 'international' minorities of all types. This conception of the nation-state was systemised and justified following the defeat of 1870 by Ernest Renan who defined the nation in terms of a will to live together, *le vouloir vivre collectif*. But the Republic which emerged from the military débâcle, the collapse of the Second Empire and monarchical divisions was fully conscious of the weakness of the social consensus supporting the new institutions. Although the Constitution could proclaim that 'the republican form of government cannot be subjected to revision', this incantation could only exorcise fears of a *coup d'état* or a change of governing majority. The primary task of the new Republic would be to create this *vouloir vivre collectif* by assimilating the marginalised sections of the population and spreading those values established by the centre. This transformation of 'peasants into Frenchmen'[6] was carried out principally by the school and the army. Indeed, the two instruments were used together: the mobilisation of the Breton reserves in 1870 was rejected because they were suspected of monarchist sympathies and of being unable to understand the orders of the French officers.

The attempt to assimilate linguistic and regional minorities would be pursued more or less successfully between the two world wars with the support of those parts of the regional population most favourable to integration: the commercial and industrial bourgeoisie, Christian Democrats favourable to the Republic and hostile to the reactionary aristocracy, the regional press, civil servants from those regions where the administration provided the source of recruitment for the most educated part of the population. But after the Second World War, the state apparatus was radicalised in reaction to the compromise made by certain regionalist militants with the Nazi occupation. More than ever, the creation of the one and indivisible French republic was on the political agenda. In 1946, the debates of the Constituent National Assembly revealed the almost total unanimity of the political world in support of the centralist conception of the state. The Socialists of the SFIO glorified the *département* as the 'crucible in which all provincial particularities are dissolved to create the cement of French unity', while Maurice Thorez declared that, for his part, and in the name of the Communist Party, 'I am opposed to any federalist ideas and I support the principle of a one and indivisible Republic'.[7] Somewhat later, hostility to the cautious regionalisation policy of Vichy would lead to the suppression of all the regional levels of the state administration.

Several more years would be required before the regionalist and federalist movements regained some semblance of respectability and were able to reassert the specific identity of their regions. This identity would be expressed through the declaration that cultural and ethnic transcend political, social and religious cleavages, the Breton movement synthesising this credo in their slogan 'Neither red nor white but Breton'. This rallying cry barely concealed, however, the weakness of those groups claiming a right to an ethnic, regional identity: in the first place, they mustered only a handful of militants who were certainly extremely active and outspoken but who failed to mobilise more than several thousand votes either in Alsace, Brittany, Corsica, or the Basque Country. In the second place, the unanimity which was proclaimed was revealed only in hostility to the centralising state, to Paris the capital. Yet these movements were rent by internecine disputes between factions which reflected the cleavages of national political organisations: moderates, those nostalgic for the traditionalist right, and extreme left-wing militants all contested the narrow political ground for the articulation of regionalist or nationalist demands.

Finally, the claim of each of these movements to be the exclusive or principal spokesman for ethnic or regional groups was confounded by the complications of national politics which created issues quite different in nature and scale: the regionalist spirit was not sufficient to withstand the impact of national electoral competition. Each election in France has represented a repudiation of the regionalist movements. The explanation given by Juan Linz for a similar phenomenon in Spain can easily be transposed to the French case: 'Many were surprised that the very vocal nationalist and regionalist parties which had occupied the limelight during the pre-democratic period in the unitary organisations of the opposition at the regional level made such a poor showing. The important role in the formation of parties in that pre-democratic period of intellectuals, professionals and students contributed to the distorted image of the intensity of nationalistic and autonomy movements and the extent to which other issues appeared more central to the voters.'[8]

The manifestations of this identity have been rather paradoxical in form. On the one hand, the specific or original character of the French regions has been increasingly diluted: local languages are in continuous decline, local customs and dress have disappeared, newspapers in regional languages have gradually lost ground, the national mass media has penetrated all regions and imposed national values, reference points and stereotypes; income inequalities and differences in lifestyle have diminished; and electoral behaviour has itself been 'nationalised' by an attenuation of the particularist behaviour of certain regions (Gaullist or Democratic Christian votes in Brittany, Alsace and the Basque Country).

On the other hand, regional integration has been continuously reinforced at the economic and political as much as the cultural level and it has been met by an equally forceful rejection by the regionalist or nationalist movements:

the more irremediable the gradual disappearance of the languages and traditional forms of minority cultures, the more strident their demands have become. The protests of intellectual and university circles, centred around analyses and slogans such as those of 'internal colonialism', were fuelled by the intolerance of the governments of the Fourth and Fifth Republic which conceded only several optional hours of teaching each week in schools and some minutes each day of television or radio broadcasting. The left in opposition adopted the regional cultural demands as its own, and criticised the Giscardian policy of 'cultural charters' as a half measure or sham. After the victory of Mitterrand, the new governing coalition facilitated the teaching of minority languages, increased television and radio broadcasting and even created a committee in Corsica responsible for dealing with these problems. This strategy has proved rather successful and has demonstrated that the problem is without doubt more political than linguistic. The concessions made by the left have served to defuse regional protests at this level and to deprive the minority groups of one of their favourite arguments as well as one of their most successful rallying causes.

In offering them what Derek Urwin has so elegantly termed 'a right to *roots*' as well as 'a right to *options*',[9] the state has laid the groundwork for a consensual integration of far greater efficacy than earlier, more authoritarian, attempts.

INTERACTION OF CENTRAL–LOCAL POLITICO–ADMINISTRATIVE ÉLITES

Sharpe has shown how various decentralised local institutions have been used 'to resist homogenizing, socio–economic forces'.[10] In addition, local or regional communities have not only been the bastion of resistance to cultural power but they have also served as the locus for the participation of social groups otherwise excluded from the political system. Until the Second World War, the geographically, economically and culturally most peripheral regions were linked to central political institutions principally through *notables*, the more-or-less faithful intermediaries of local pressures and demands. This mode of intermediation gave rise to serious conflict not only between the representatives of the values of local society and those of the coalitions in power but also within local society itself over the mastery and control of this mediating role: a struggle among clans in Corsica and between the landed aristocracy and Christian Democrats in Brittany for example.

After the Second World War, those representatives who had collaborated, to a greater or lesser extent, with the Germans found themselves disqualified from office to the advantage of the élites emerging from the Resistance. The framework in which the replacement of élites occurred, however, at least as far as centre–periphery relations were concerned, remained unchanged. In effect, the attempts of the left to limit the influence of rural *notables* and strengthen that of left-wing elected representatives almost totally failed. The left had first attempted to abolish the Senate, guilty in its eyes of having sabotaged the Popular Front of 1936. The Senate's fundamentally rural composition (whence the ironic title of 'Chamber of Agriculture') due to an electoral system which favoured the representation of the 39,000 communes of France, had, under the Third Republic, protected the urban bourgeoisie from 'an excess of universal suffrage'. But General de Gaulle opposed this radical reform with his own proposal for a representative assembly of economic and social interests, thereby preventing the abolition of the Senate. The left did succeed in having the prefects diminished in status from departmental executives to representatives of the state. However, the cold war, the hostility of the right and the resistance of the prefectoral corps combined to prevent the implementation of the reform envisaged by the Constitution. The existing system of prefect–*notable* relations, so characteristic of a centralised state dominated by its bureaucratic and *notable* élites, would remain self-perpetuating.

In the system of mediation defined by Thoenig[11] as a 'honeycomb structure' there were few alternatives to institutional channels as means for the expression of interests in civil society.

Prefects and *notables* are policy-brokers[12] assigned the role of spokesmen for both the centre and the periphery. The advantages of such a system have been presented. Its drawbacks, however, are far from negligible. Not all demands and claims can be articulated through the traditional channels. During the 1950s, protests and demands were voiced above all by those social groups most affected by economic change: small shopkeepers, artisans and peasants. These protests did not, however, assume the same forms in all regions. The most dynamic peripheral regions mobilised in favour of economic development through the intermediary of associations composed of those economic and social interests likely to benefit from 'economic expansion'. The 'expansion committees' in regions or *départements* provided the organisational expression for a form of protest which found its first spokesman in Jean-François Gravier. A number of such committees were created during the 1950s, the most notable of which was the Breton Committee for Studies and Interest Group Liaison (CELIB).

The objective of these committees was to mobilise interests (later to be called *forces vives*) in favour of economic development by seeking to transcend the traditional cleavages of French society. Trades unions and employers, academics and peasants, regionalists and economic leaders, parties of the left and right, were all invited to rally to the cause of regional development. With the exception of the Communist Party – which was highly suspicious of this kind of unitarian project – practically all groups, associations and parties were presented on these committees. Their intention was not only to encourage regional initiatives but to function as a powerful pressure group capable of influencing the decisions of central government either directly, or through the intermediary of the elected representatives of the regions who were constantly reminded of their mediating role between centre and periphery.

In certain areas, the protests of disadvantaged sections of the population took a corporatist rather than regionalist form. The Poujadist movement between 1954 and 1956 was the most powerful and vocal of those movements representing professions in decline. Together, these two types of movement constituted a formidable threat to central power and to the position of the ruling political class. Pierre Pfimlin, a Christian Democrat from Alsace, could, of course, evoke their positive aspects: 'This is a remarkable phenomenon', he declared before the National Assembly on 6 August 1954, 'this spontaneous emergence in thirty or forty regions of France of men of good will who, in one way or another, have dedicated themselves to the task of reviving their region or *département*'. In 1954–55, the governments of Mendès-France and Edgar Faure responded very quickly to the regional demands, in an attempt to contain pressures from the periphery which threatened to become uncontrollable. The method employed was one of the traditional weapons of the French administrative arsenal: dialogue with the interest groups concerned, but with the state reserving the right to nominate its negotiating partners and to fix the rules of the game. It was thus that the decree of 11 December 1954 'authorised'(!) the creation of regional or departmental 'committees of economic expansion'. Only those committees 'approved' according to criteria of representation defined by the government could be 'consulted on those measures pertaining to local economic development in the government's general policy framework'. Other committees could continue in existence but only those approved would enjoy the legitimacy (and advantages) conferred by governmental decree.

The arrival of the Gaullists in power in 1958 led both to the decline of the expansion committees and to their manipulation to the benefit of the central power. Gaullist elected representatives refused henceforth to favour their regional allegiances at the expense of the partisan discipline imposed by the centre. Partisan cleavages would henceforth be given free expression and bring an end to the period of often artificial unanimity among regional groups. As the committees were progressively weakened, they could be taken in hand by central government and deployed against the local representatives who, for the most part, were hostile to the power of the centre. The Gaullists now attempted to find local level allies who could help counter the resistance of left-wing and centre notables. The electoral reform of 1965 in towns of more than 30,000 inhabitants provides an example of this strategy.

It was with the creation of the CODERs (committees for regional economic development) that the government hoped to supplant the traditional political élites with new élites from socio-professional groups described as *forces vives*. This attempt had the full support of the senior civil service which sought to develop local networks of support and the means of legitimation for its policy of planning and development. The most ardent supporters of a policy of participation at the time were those administrative élites which, from Gaullist circles, through the moderates of the Club Jean Moulin, to the extreme left, aimed to offset the extreme concentration of power and find alternatives to dialogue with discredited groups. 'Participation' was to be the magic word of 1968, and de Gaulle believed he had discovered an adequate response to this demand in his proposal to encourage socio-professional representation in the regions and within the Senate. The failure of the 1969 referendum confirmed the victory of the *notables* and destroyed any dreams of replacing local élites by Gaullists. This replacement would occur during the 1970s, but would work rather to the advantage of the left which succeeded in conquering first the periphery and subsequently the centre.

PATTERNS OF CENTRAL INTERVENTION

The penetration and control of the periphery have been and remain the constant preoccupation of most European states. The forms taken by penetration are varied and usually rather unsophisticated: military penetration; strict control of central power over the periphery; domination of central élites; economic integration; cultural imperialism. Studies in the 1960s tried to demonstrate — notably through the thesis of 'internal colonialism' — that the mechanisms of penetration had been economic in form as well as political and cultural. Robert Lafont[13] was the principal exponent of this school in France. This approach enjoyed an undeniable success, particularly as an instrument of political mobilisation: the mass media and regionalist pressure groups rallied *en masse* behind the slogan 'Decolonise the regions', reinforcing the impression that penetration was unilateral and to the sole advantage of the state apparatus. In reality, as the studies of the Centre for the Sociology of Organisations had shown, this process may more accurately be characterised as one of *interpenetration* rather than unilateral penetration by the centre.

A noteworthy example of this can be found in the phenomenon of accumulation of offices (*cumul des mandats*), one of the most enduring 'conventions' of the French political system. After the victory of the left, far from having diminished, this phenomenon increased in importance: 93 per cent of Senators and 82 per cent of Deputies held at least one local office in 1983. This can only strengthen the concentration of power, the oligarchy within élites and the osmosis between central élites and local élites. This interpenetration is not only encouraged but was sometimes compulsory, as in the case of regional councils of which local Deputies and Senators were obligatory members up to 1986. Despite the contradictions and conflicts inherent in the relationship between centre and periphery, the accumulation of offices helps to ensure a symbiosis and convergence of interests which breaks down only if pressures and social protest are exceptionally strong. As Becquart-Leclerq has pointed out, 'the central and local logics thus converge on certain key-nodal positions for the practice of the power of intervention. This being the case, the accumulation of offices serves to strengthen the framework of relational linkages and to ensure system stability while rendering far-reaching reforms which upset too many interrelated interests extremely difficult to achieve'.[14] It was not until 1985 that the Socialist government succeeded in enacting a bill which limited the number of offices to two. However the legislation does not cover offices in cities of less than 20,000 and, consequently, the impact of the law is restricted because only 900 communes (out of 36,000) have more than 20,000 habitants.

In spite of their 'complicity' with the central government, the *notables* had to take into account in the 1960s and 1970s the pressures from the periphery. The upsurge of 'nationalities' within France made itself felt within the political parties. While rejecting the autonomist parties and criticising their violent methods, the Communist Party called for respect for regional languages in schools, in the press and in radio and television. The Socialist Party, which had integrated a section of the PSU, created a 'commission for national minorities' and attempted thereby to enlarge its electoral appeal in those regions which were still incompletely integrated, such as Brittany. The Radical Party, with its traditional base in Provence and Corsica, adopted some of the regionalist demands as its own.[15] Finally, the parties of the right-wing coalition made a number of concessions, most notable among them the policy of Cultural Charters.

This success of the autonomist and regionalist movements was, however, more apparent than real, and political developments in recent years have revealed their failure. There have been various reasons for this. In the first place, the integration of the regional *problématique* into the national political arena has limited its radical scope. Political parties have proclaimed constantly 'Let us regionalise, let us regionalise' but without mobilising themselves effectively in support of such reform. The electoral process had done the rest: owing to the national character of electoral campaigns, the regional issue has been only one among many, and often secondary to others, in the political struggle. Those regionalists who have dared to venture on to the electoral stage have often done so at their own expense: the Breton Democratic Union chose this strategy (in preference to the bombs which are the prerogative of the Breton Liberation Front) and was poorly rewarded in return; less than 2 per cent of the vote in the 1978 legislative elections, a few dozen (36) seats in the 1977 local elections thanks to its alliance with the Union of the Left. In the 1986 regional elections (the first by direct suffrage) the autonomist movements won only six seats out of 1800, four of them in Corsica!

Secondly, the response of the state has successfully prevented the transformation of isolated minority movements into a vast regional coalition. The government has been able to repress autonomist activists with the tacit support of national political figures opposed to 'any violence against property or persons'. For many years, Breton, Corsican, Basque, West Indian and Tahitian militants were brought before the state *Cour de Sûreté* and condemned to heavy or light sentences depending on the strategy pursued by the government at the time. But at the same time, the guardian state has looked down upon its prodigal sons and provided the financial manna required for the eradication of the evil. The unruliness of the Bretons and Corsicans has long earned them the solicitude of state power, and continues to do so. Overseas, the situation has become absurd: the standard of living of the West Indians and Tahitians depends so heavily on the metropolitan welfare system that the autonomist demands have lost a great deal of their credibility.

Thirdly, those transitory factors which worked in favour of regional consciousness 20 years ago now work against regionalisation and local autonomy. The international crisis has affected all nations, but within them, the weakest areas (rural regions and industrial regions in decline) have suffered most from its impact.

Finally, the importance of the phenomenon of regional culture expressed in songs, music, literature and local languages was overestimated and, ultimately, undermined by the commercial networks of the consumer society. There are certainly more students learning Breton or Provençal but these languages have withered on stony ground, find no place for expression in the press and have access to only several minutes of television broadcasting each week. Folklore has had its moment of glory but has faded before the changing demands of show-business.

ECONOMIC REDISTRIBUTION

Although emphasis has often been placed on the ethnic or linguistic aspects of peripheral demands, the phenomenon cannot be reduced to cultural problems, nor can they be considered the most important element of the revolts and protests. In fact, one of the major features of regionalism since the Second World War has been its radical transformation. The periphery has not mobilised first and foremost for its culture but rather for its economic development. Even in those regions with a strong cultural character such as Brittany or Corsica, regionalist demands for a better territorial distribution of economic development have overriden cultural considerations. The leaders of regional mobilisation have been very different from those of the nineteenth century and the inter-war period: the landed aristocracy and academics specialising in local history have given way to unionised peasants or workers, academics concerned with regional planning and development (economists and geographers) and the officials of Chambers of Commerce.

Paradoxically, it is the increasing integration of the regional economies within the capitalist system which has spurred the revival of regionalism. The price to be paid for economic growth by numerous social groups has become clear: peasants exiled to the towns, rural artisan industry condemned to disappear by the modern methods of production, agricultural workers driven from the land by mechanisation, small shopkeepers and industrialists ruined by new modes of production and distribution, intellectuals incapable of finding employment in their regions of origin and forced to emigrate to large urban centres. As noted above, the protests of social classes 'stricken by progress' have often been expressed in sectoral fashion, but intellectuals (in the regions or, more often than not, emigrants to Paris) have given, with the help of their theories, a common denominator to the specific problems of the affected social groups. Renaud Dulong has pointed out how the territorial framework which might only have been an empty shell became the means by which social actors attributed a wider significance to issues which otherwise could not have been articulated.[16]

This 'territorialisation' of economic and social issues has had one rather paradoxical consequence. The policy of economic regionalisation which was called for in the name of *regionalism* has led both to increasing state intervention and to the economic integration of the regions. This contradiction which, in general, has not been appreciated by the regionalist movements explains why the central state has often been so favourable to a policy of regional development and planning when its attitude to cultural or political issues has more often than not been one of reserve. Brittany, Corsica and, above all, overseas territories have benefited from the financial generosity of the state. But it was not until 1981 that a number of concrete, if far from revolutionary, measures were taken in response to the demands for regional autonomy. On the other hand, the policy of regional planning and development has had the advantage of seeking to reconcile the conflicting interests of Paris and the regions. The rhetoric of industrial decentralisation should not conceal the fact that 50 per cent of the 350,000 jobs 'decentralised' from 1950 to 1964 were restricted to a radius of 200 km from Paris, that 'the brake' on Parisian expansion was compensated for by the creation of several new towns just outside the boundaries of the capital. And even the decentralising reforms of the left should not disguise the considerable centralisation of economic power involved in the nationalisation of the large industrial groups and 36 banks.

It is clear that the economic crisis has changed the mind of governments. In the 1950s and 1960s, the central government answered the regional challenge by redistributing economic activities towards the poorest areas. It pursued an active policy of *'aménagement du territoire'*, notably through an *ad hoc* inter-ministerial agency created in 1964, the *DATAR (Délégation à l'aménagement du territoire et à l'action régionale)*. Today, the DATAR still exists but its capacity to direct investments, public or private, towards the more underdeveloped regions has weakened. The policy of the day is much more undifferentiated, and locational problems are considered less important than before. Economic growth and industrial development are the key words and have supplanted earlier priorities, favouring balanced distribution of activities over the national territory. It is revealing that the very existence of a body such as the DATAR has been called into question.

CONCLUSION: INCREMENTAL ADAPTATION AND DEMOCRATIC INTEGRATION

While remaining faithful to its unitary and Jacobin principles, the left was aware that an integration founded on consensus could be far more cohesive than one achieved by the more traditional, authoritarian methods. The new legitimacy of the state, based on a recognition of groups and of local and regional values, found its expression in the strengthening of direct elections and the symbolic weakening of the power of the state. This hardly amounts to a revolution; but it did signify an adaptation to the demands of territorial democracy. Indeed the 1981–86 Socialist reform programme was not justified in terms of public managment – even if the problems of the division of labour, of laws and of personnel management played an important role. In the eyes of the left, the local system, quite independently of its defects, suffered from two main problems: a lack of democracy in its institutions and the arbitrary character of state intervention. *Changer la vie, changer l'Etat* meant that, at the local level, priority should be given to changes considered indispensable for justifying the activity of local élites. In other words, even before modifying or extending the powers to be devolved to the various decentralised levels, the left considered it essential to invest it with a new legitimacy. In this respect the beliefs of the left are well known and have been continuously reasserted during two centuries of political struggle: election is the source of all power, and the rule of law is protection against the arbitrary. The left (Socialist Party, Communist Party, Radical Party) have, since the last war, persistently demanded the 'democratisation' of the local level and the weakening of central control. The Constitution of the Fourth Republic, which was considerably influenced by the left, was the first in French history to express this political inspiration, but it was never put into practice. It was, therefore, no surprise that the reforms of 2 March 1981 gave formal recognition to the democratic ideals repeatedly voiced by the left.

The legitimacy of power is henceforth assured at all local levels by recourse to election: local, departmental and regional assemblies are all directly elected and they appoint their own executives.

The innovation involved in this general recourse to election should not obscure the fact that other mechanisms for democratic accountability were not considered: referendums, public hearings, neighbourhood councils. More generally, all the other means of creating a participatory and responsive democracy were ignored. One might welcome this oversight because these options are less than perfect: they are too often manipulated by unrepresentative pressure groups and they can be administratively disruptive. But it must be said that the existing channels for expressing opinions and applying pressure in society are singularly limited. 'One of the unwritten principles of French public law is that people are incapable of self-government', and Céline Wiener has added, 'Citizens are merely able to choose good representatives, but they have no direct role, simply that of electing delegates who will manage or control the managers in their names'.[18]

The inadequacy of the Socialist reforms is forcefully illustrated by the fate of urban associations, the ecologists and the extreme left – in short, by all those whom the politico-administrative system tries more or less to marginalise. As early as the 8 August 1981, for example, the Trotskyist *Lutte Ouvrière* expressed its disappointment by claiming that the decentralisation proposals would effect only a modest transfer of powers from the centre to the prefects and from the prefects to the electors. But 'for the population as a whole, the changes will by and large be invisible'. Furthermore, local authorities will henceforth be subject to a system of legal controls as in most other Western countries – that is, they will be answerable to a *judge*. The new organisation or model of centre–periphery relations follows, at least at the formal level, the example of the majority of West European countries. The experience of these countries leads us to expect at least two sorts of difficulties in putting such a new system into operation. On the one hand, there is the possibility of inflexibility in the exercise of jurisdictional control compared with the margin for manoeuvre and negotiation which always characterised prefectoral supervision. On the other hand, there is a considerable risk of a politically motivated supervision by one tier of local government over another, especially in the fields of planning controls and subsidies for investment projects. These possibilities represent very real fears for numerous mayors, especially those of small communes. Something similar has already occurred in Italy where municipal and provincial elected representatives complain *mezza voce* that elected regional representatives have replaced the prefect in exercising supervision.

This process of change does not, however, go so far as to transform the fundamental characteristics of the French politico-administrative system. One of the most significant features of the French local government system is the personalisation – indeed, the authoritarianism – which characterises the exercise of power: the length of time that people hold an elected office, the fact that many elected posts are handed on from father to son, the multiplicity of electoral offices held by a single individual, the methods of decision-making are just some of the manifestations of a phenomenon which has little or no equivalent in the other countries of Western Europe.[19] This model has until now had two variants: the prefectoral and the mayoral. The first has been abolished, but otherwise the model remains intact. The prefects have been replaced by elected executives, but the concept of a strong power remains — as shown by the rejection of the Communist Party's proposals for a more collegial system of decision-making. It is not our task here to assess the (numerous) advantages or the (equally numerous) inconveniences of the PCF's proposals. The general point remains that the change in the élites has not resulted in any substantial modification in the way things are run.

Perversely, the new system is open to exploitation by local and national élites. As a result, the problems of 'government by delegation' may well be exacerbated. The élites continue to hold multiple offices and, since they cannot exercise all of them effectively, they will continue to have recourse to *éminences grises*. The latter are the real managers of day-to-day affairs (and often more) since the mayor, the chairman of the *conseil général* and the regional council are simply unable, through lack of time, to accomplish all their tasks, any one of which would take up most of their time. Well-staffed private offices (*cabinets*), which have already been in existence in the big towns for some years, are now being introduced at departmental and regional level. Already about 60 prefects and sub-prefects have been seconded from the prefectoral corps to run things in *départements* or regions on behalf of the new elected representatives.

Thus, a strange game of 'musical chairs' is being played, with one part of the old prefectoral corps removed but reappearing in other guises. And the local system is run by public servants to a greater extent than before. Already numerous observers have identified the degree to which public servants have assumed elected office at local and national level in the last few years. But the phenomenon is even more striking if one looks at what has happened in the most important areas. Overall, about 10 per cent of the mayors are also public officials. However, the proportion is considerably greater in larger and more urbanised towns. For example, Mabileau and Sadran have shown that in the region of Aquitaine, more than half the mayors of towns with more than 10,000 inhabitants are also public officials. They also note that 'the higher you rise in the hierarchy of local political responsibilities, the larger the proportion of public servants' (26 per cent of town councillors, 32 per cent of assistant mayors and 52 per cent of mayors). Similarly, whereas less than 20 per cent of departmental councillors are public officials, amongst the members of the *département* executives, this proportion rises to more than a quarter.[20]

We are witnessing, therefore, the take-over of local government posts by public servants by two means: *sociologically* in the candidates elected, and *functionally* by the growing practice of public officials carrying out the administrative and policy tasks of elected individuals. Incidentally, the reform of the local public service has also provided a solution (of sorts) to a problem of legitimacy: local public servants are now subject to rules governing recruitment and career similar to those applying to state civil servants. The time when elected representatives could recruit just whom they wanted *selon leur bon plaisir* seems to have disappeared — at least in principle.

NOTES

1. J. Gottman, 'The Evolution of the Concept of Territory', *Social Science Information*, 14 (1975), p. 29.
2. J. F. Gravier, *Paris et le désert français* (Paris: Flammarion, 1947).
3. R. Aqua and R. Gates, paper to the conference on *Local Institutions in National Development*, Bellagio (1982), p. 1.
4. S. Huntington, *Political Order in Changing Societies* (New Haven: Yale University Press, 1968), p. 93.
5. S. M. Lipset, 'The Revolt against Modernity', in P. Torsvik (ed.), *Mobilization, Centre-Periphery Structures and Nation Building* (Bergen-Oslo, 1981), p. 319.
6. E. Weber, *Peasants into Frenchmen* (London: Chatto & Windus, 1976).
7. Journal Officiel, Débats, Assemblée Nationale, 22 March 1946, pp. 988 and 991.
8. J. Linz, 'Party Systems in the Periphery', paper presented at the *Conference on Recent Changes in European Party Systems*, European University Institute, Florence, December 1978, p. 7.
9. D. Urwin, 'The Price of a Kingdom: Territory, Identity and the Centre-Periphery Dimension in Western Europe', in Y. Mény and V. Wright, *Centre-Periphery Relations in Western Europe*, (London: Allen & Unwin, 1985), pp. 151–70.
10. L. J. Sharpe, 'Decentralist Trends in Western Democracies: A First Appraisal' in L. J. Sharpe (ed.), *Decentralist Trends in Western Democracies* (London and Beverley Hills: Sage Publications, 1981), pp. 9–79.
11. Jean-Claude Thoenig, 'State Bureaucracies and Local Government in France' in K. Hanf and F. W. Scharpf (eds.), *Interorganizational Policy Making* (London: Sage 1978), pp. 184–6.
12. S. Tarrow, *Between Center and Periphery: Grassroots Politicians in Italy and France* (New Haven and London: Yale Univesity Press, 1977).
13. R. Lafont, *Décoloniser la France* (Paris: Gallimard, 1971).
14. J. Becquart-Leclercq, 'Cumul des mandats et culture politique', in A. Mabileau (ed.), *Les pouvoirs locaux à l'épreuve de la décentralisation* (Paris: Pedone, 1983), pp. 207–41.
15. Y. Mény, *Partis politiques et Décentralisation*, (Paris: Cahiers de l'Institut Français des Sciences administratives, Cujas 1979); H. Machin, 'All Jacobins Now? The Growing Hostility to Local Government Reform', *West European Politics*, Vol. 1, No. 3 (1978), pp. 133–50.
16. R. Dulong, *La question bretonne* (Paris: Armand Colin, 1975).
17. M. Bouissou, 'La pratique référendaire en France', *Revue Internationale de Droit Comparé*, (1976), No. 2, pp. 270–86.
18. Céline Wiener, 'Service public ou autogestion: d'un mythe à l'autre', *Mélanges Charlier*, (Paris: Ed. de l'Université, 1981), pp. 325–40.
19. Y. Mény, 'Le maire, ici et ailleurs', *Pouvoirs*, No. 24, 1983, pp. 19–29.
20. A. Mabileau and P. Sadran, 'Administratio et politique au niveau local', in J. L. Quermonne (ed.), *Administration et politique en France sous la Ve République* (Paris: Presses de la FNSP, 1981), pp. 257–87.

The Federal Republic of Germany: From Co-operative Federalism to Joint Policy-Making

Joachim Jens Hesse

The most important development in West German federalism over the first three decades after the Second World War has been the increasing interdependence of the several levels of government. Political scientists in the Federal Republic of Germany have paid special attention to this subject, producing numerous case studies under the keyword *Politikverflechtung* (joint policy-making)[1] and developing a new theoretical perspective on the study of intergovernmental relations. Federalism was said to be characterised by a deterioration of governability in systems of joint decision-making, a general decline of problem-solving capacities, and decreasing efficiency and innovativeness.

There is no doubt that *'Politikverflechtung'* was an important new approach. It introduced political and bureaucratic factors into a traditional conceptualisation of intergovernmental relations which scarcely considered conflicting interests, processes of co-operation and bargaining, and problems of control and implementation. Yet, in comparison with other West European countries, the intergovernmental system of the West German state has proved to be surprisingly stable and adaptable to changed conditions. Consequently, the theoretical framework developed during the 1970s seemed to be incomplete. Restricted to case studies of specific policy areas, it failed to provide a comprehensive and dynamic approach for understanding developments in intergovernmental relations over a longer period.

This study starts from the assumption that the relative stability of West German federalism is rooted in the special adaptability of the system; an adaptability which has not been considered in the discussion on joint policy-making. The structure of the West German intergovernmental system is characterised by an extensive entanglement and interdependence between levels of government and by co-ordination and co-operation among federal, *Länder*, and local governments, which might cause – in the short run – inadequate, suboptimal policy outcomes. However, a longitudinal analysis of intergovernmental relations reveals the considerable ability of the West German state to react and adapt to changing economic and socio-cultural conditions. To justify and expand on this hypothesis, we provide – after a short description of the institutional setting (section 2) – an analytical framework which explicitly focuses on the role of the municipalities and the influence of changing socio-economic conditions on intergovernmental structures and processes (section 3). The development of the intergovernmental system in Germany since 1945 is described in order to demonstrate the complexity of the structure and, above all, the incorporation of the local level (section 4). Finally, current developments in central-local relationships are analysed in order to assess the effects of trends towards decentralisation and fragmentation of political power in the West German federal system (section 5).

INTERGOVERNMENTAL STRUCTURES IN WEST GERMANY

When the Constitution of the Federal Republic of Germany was discussed and finally adopted in the Parliamentary Council (*Parlamentarischer Rat*), the advocates of a decentralised federal state with far-reaching autonomy for *Länder* and municipalities were in the majority. In contrast to traditional federal systems, well-developed forms of co-operation and of joint functions, decisions and resources were incorporated in the construction of the West German federal state. In the course of developments since 1949 this 'co-operative federalism' has been extended, and the tendency for all territorial levels to become interlinked has intensified.

Today, the federal state of West Germany is characterised by a sharing of functions, which confers the most important fields of legislation on the federal government, whereas the *Länder* are responsible for the implementation of federal laws. Although the Basic Law stipulates the supremacy of the *Länder* even in legislation, the federal government has, over the years, extended its legislative functions, either by an extensive use of concurrent legislation, or by amendments to the Constitution agreed by the *Länder*. The federal government has also extended its executive functions via the growth of the federal administration. These trends toward centralisation and unitarisation even affected education and police services, the most important functions exclusively reserved to the *Länder*. The 1969 constitutional reforms introduced so-called joint tasks (*Gemeinschaftsaufgaben*) in some areas of responsibility. Since then, the construction of universities (including university clinics), the improvement of the regional economic and the agrarian structures and coastal preservation have been jointly planned and financed by the federal government and the *Länder* on the basis of skeleton provisions (*Rahmengesetze*).

From the legal viewpoint, the municipalities are part of the *Länder*. In principle, they have the right to regulate all the affairs of the local community. The scope of their responsibilities is, however, a matter of some dispute. Arguments focus on whether or not an area of exclusively local functions can still be delimited. The municipalities and the counties (*Kreise*) perform essential tasks, including infrastructural development, cultural affairs, public utilities and social welfare. They are responsible for approximately two thirds of overall public investment. Furthermore, the counties and municipalities perform delegated state functions (*Auftragsangelegenheiten*), which considerably increased during the 1950s and 1960s.

The entanglement of levels of government in the West German state can be demonstrated most clearly by looking at the structures and procedures for intergovernmental decision-making. The Basic Law contains provisions which foster intense co-operation between federal and *Länder* governments. In the upper House of Parliament (*Bundesrat*) representatives of the *Länder* governments participate in federal legislation, and a majority of votes in the *Bundesrat* can even — by right of veto — prevent the adoption of bills which interfere with the interests of the *Länder*. Since this provision applies when a bill

TABLE 1
DECISION-MAKING STRUCTURE IN THE WEST GERMAN INTERGOVERNMENTAL SYSTEM

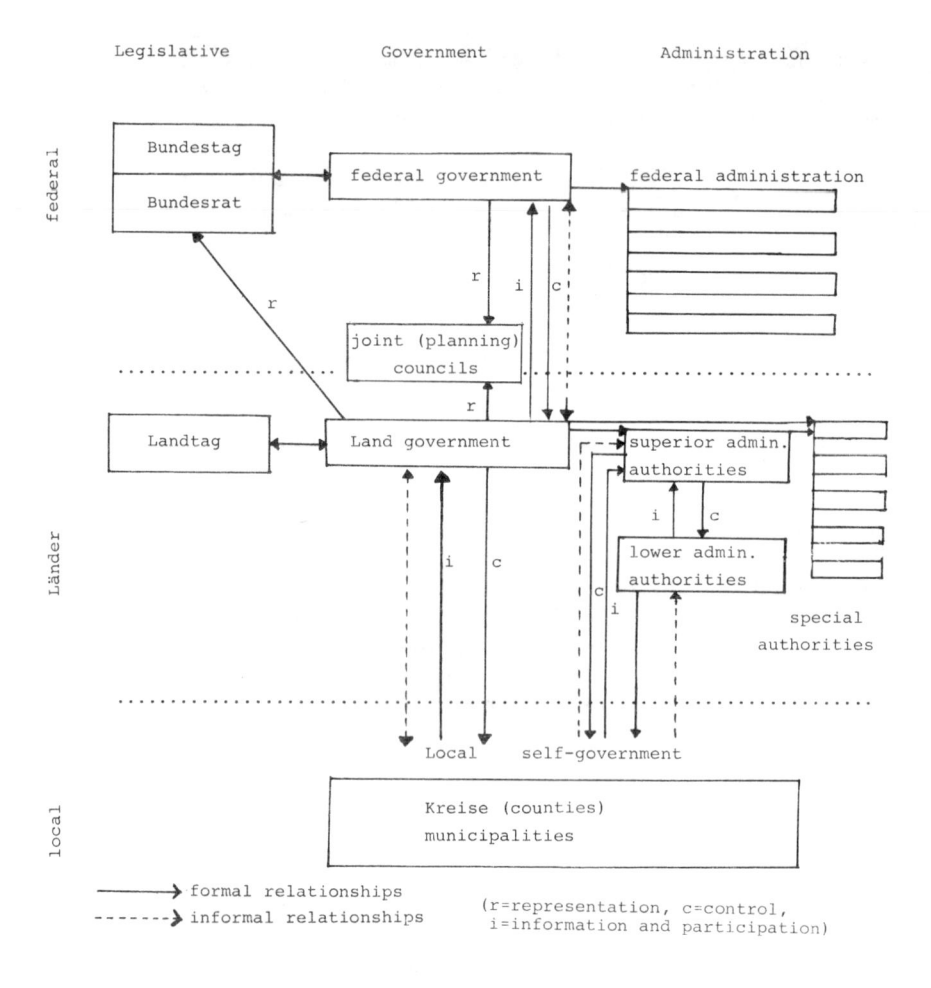

```
                  Legislative         Government              Administration

  federal     ┌──────────────┐   ┌──────────────────┐    ┌──────────────────────┐
              │  Bundestag   │◄──┤federal government├───►│federal administration│
              ├──────────────┤   └──────────────────┘    └──────────────────────┘
              │  Bundesrat   │           ▲
              └──────────────┘        r  │ i  c
                      ▲        ┌──────────────────┐
                      │        │ joint (planning) │
                      │        │     councils     │
                      │        └──────────────────┘
                      │                r │
              ┌──────────────┐   ┌──────────────────┐    ┌──────────────────┐
  Länder      │   Landtag    │◄──┤  Land government ├───►│  superior admin. │
              └──────────────┘   └──────────────────┘    │    authorities   │
                                                         └──────────────────┘
                                       i │ c                    i │ c
                                                         ┌──────────────────┐
                                                         │   lower admin.   │
                                                         │    authorities   │
                                                         └──────────────────┘
                                                                               special
                                                                             authorities
                                       Local   self-government
                                      ┌──────────────────────┐
  local                               │   Kreise (counties)  │
                                      │    municipalities    │
                                      └──────────────────────┘

  ─────► formal relationships
  ─ ─ ─► informal relationships    (r=representation, c=control,
                                    i=information and participation)
```

stipulates how a policy is to be implemented by the *Länder*, federal legislation depends on the approval of the majority of *Länder* representatives in a number of substantive policy areas.

It is also important to note that the higher territorial authorities (*höhere Gebietskörperschaften*) exercise supervisory control over the lower level authorities. As a result, there are many significant opportunities for *Länder* governments to influence local policies. Yet the lower level authorities have a right to participate, to enjoy access to information, whenever the higher authorities interfere in their jurisdiction in matters of planning and control. It is hardly surprising that this situation led to an intensification of bargaining relations between the different levels of government.

Fiscal relations have changed several times since 1949. Initially, the Constitution attempted a clear demarcation of financial sources for the different administrative authorities. The constitutional changes between 1953 and 1956 sacrificed this principle, because of the increasing gap between revenue and financial needs at the several levels of government. The economic upswing of the 1950s especially favoured the federal government, which profited most from the high tax revenues,[2] whereas the capital-intensive functions of investment and administration were primarily discharged by the *Länder* and municipalities. To compensate for these divergent trends, the federal government allocated further subsidies.

The 1969 financial reform extended resource sharing. Since 1969 the majority of taxes have been shared or joint taxes (*Gemeinschaftssteuern*). The apportionment is as follows:

- Income Taxes: federal government and *Länder* 42.5 per cent each, municipalities 15 per cent.
- Corporate Profit Taxes: federal government and *Länder* 50 per cent each.
- Value Added Taxes: federal government 65 per cent, *Länder* 35 per cent in 1986; apportionment redetermined annually.
- Business Taxes: federal government and *Länder* 20 per cent each, municipalities 60 per cent (since 1979 the share of the federal and *Länder* governments has been reduced repeatedly and amounts now to about 8 per cent each, leaving 84 per cent to local governments).

Tax revenues are distributed through a complicated system of fiscal transfers (*Finanzausgleich*) between federal government and *Länder* (particularly federal supplements to financially weak *Länder*), between the different *Länder*; and between the *Land* and municipalities. In addition, grants are given by the federal government to the *Länder* or by the *Länder* to the municipalities. Federal grants to the *Länder* form about 17 per cent of their revenues. Local governments receive 41 per cent of their fiscal resources in the form of grants (8.7 per cent directly and 25 per cent indirectly from the federal government). About 50 per cent of the grants are categorical.[3]

AN ANALYTICAL APPROACH: FROM 'POLITIKVERFLECHTUNG' TO 'INTERGOVERNMENTAL DYNAMICS'

The normal perception of West German federalism received a severe blow in the mid-1970s, with Scharpf *et al.*'s review[4] of the problem-solving capacities of the intergovernmental system. Focusing on joint tasks (*Gemeinschaftsaufgaben*) and grants (*Finanzhilfen*) this review demonstrated the weakness of traditional analyses of federalism. The legal or economic perspective, with its marked normative component, tended to omit crucial bureaucratic and political factors. Scharpf's theoretical and empirical analysis of co-operative federalism demonstrated that the patterns of joint decision-making stressed conflict avoidance and, at best, allowed the stabilisation of zero sum-games between the *Länder*, creating 'joint decision traps' which did not efficiently resolve redistributive problems.

Although productive, this approach cannot be generalised – in spite of its recent application to the supra-national level of government:[5] it is based on a cumulative analysis of special cases, restricted to the co-ordination of sectorally defined policies, to federal (*Bund*)-state (*Länder*) relations, and to a limited period of time. The approach has to be extended to include explicitly the local level; the socio-economic interests influencing the different territorial units; and, to facilitate a dynamic approach to intergovernmental relations, a longitudinal dimension or the long-term observation of different (routine and non-routine) processes of policy-making. The local level should be included because German municipalities may exercise considerable influence over the processes of policy-making at the federal and the state level: they initiate public policies, implement state programmes, act as an information resource for state planning, and are a focus of political mobilisation.

In this context, it will be possible to provide only a brief outline of an extended theoretical framework. The following summary draws on a series of case-studies by Fürst, Hesse and Richter and a comparative research project currently under way, by Benz and Hesse.[6]

Historically, since industrialisation, cities and the central government in Germany have had an antagonistic relationship: cities were at the forefront of the political fight for civil rights and liberties against the authoritarian state. Later, they acted to attenuate the social impact and consequences of industrialisation. In the post-industrial era the cities are to the fore in a new conflict, this time between locally expressed 'reproductive interests' (*reproduktionsorientierte Interessen*) and 'production-oriented interests' (*produktionsorientierte Interessen*), which are most strongly represented at the federal and state level.[7] 'Reproductive interests' concern the satisfaction of individual needs. They are not restricted to questions of income levels and distribution: indeed, they are partly opposed to economic interests, seeking to avoid the social and environmental costs of private production. There is also a desire to enhance individual development and to promote self-fulfilment and the 'humanisation of labour'. These terms do not represent a return to a class-theoretical analysis. The phrase 'related to reproduction' refers rather to a new set of social values which conflict with 'production-oriented interests' and generate 'role conflicts', so to speak, between, for example, an individual's membership of a trade union and his concern with the quality of life.

Increasingly, reproductive interests have been expressed in post-industrial societies, but such demands can be satisfied by market processes only to a small extent. They are articulated through the political-administrative process; a process which is dominated by 'production-oriented interests' – i.e., the government interest in revenues, the employers' interest in the exploitation of capital, and the labour unions' interest in occupational stability. 'Reproductive interests' have elicited greater response at the local level. Such a response is most likely when, for instance:

- municipalities have a relatively low dependence on resources drawn from production,

- the reproductive interests find it difficult to organise at regional or national level,
- the 'spill-over' problems of production have a disproportionate effect at the local level (e.g. pollution).

Vice versa, 'production-oriented interests' operate at the *Länder* and the federal level because of their concern with the control of economic development, the provision of infrastructure and the social policies relevant to production (for example, labour market and educational policies). It is possible to contend, therefore, that there exists a kind of functional division of labour between local and central government, especially as the local level becomes less significant for economic interests with the internationalisation of production processes. As a result, social conflicts in the nation are absorbed, not only by sectoral segmentation but in part by the functional division of labour within the intergovernmental system.

This divergence of socio-economic interests relates to the distribution of administrative interests. Thus municipalities have relinquished part of their autonomy to higher state authorities, while at the same time the latter have an interest in deconcentrating their functions to the municipalities. It is plausible to assume that the latent conflicts between 'reproductive' and 'production-oriented' interests would be neutralised by the latter, represented by the state which seeks to incorporate the local level into national problem-solving and thereby 'disciplining' it. But quite the opposite way of regulating the conflicts is possible, too: the local level enters into conflict with the *Land* and seeks to satisfy 'reproductive interests' through processes of redistribution between the different levels of government. Moreover, the *Länder* and federal government cannot ignore 'reproductive interests'. One way of minimising their disruptive potential is to strengthen the local level and to confine the problems to the locality. Moreover, even if the politicisation of 'reproductive interests' remains low at the local level the municipalities might suffer if dissatisfied interests 'voted with their feet' and left the area, with the attendant loss of revenue.

The cities can attempt to solve their problems by seeking to redistribute resources between levels of government. This redistribution can be effected in two ways. First, the cities can externalise local problems by shifting the burden on to the *Land* and federal departments, putting the *Land* under pressure to react and re-structure the definition of the problem, and at the same time instituting vertical *'Politikverflechtung'* (joint-policy making) as a pattern. Second, the cities can try to influence the distribution of resources between territorial authorities and thus change the conditions for policy-formulation. The more interesting case of the two is that of cities influencing the redistribution of capacities for problem-solving between the different levels of government, because this indicates an increase in the power of the cities in the federal system not only in special policy areas, but also in their relations with the other territorial units.

The redistribution of problem-solving capacity involves the redistribution of territorial or administrative functions. The latter may increase the former. An extension of territory can offset losses caused by exodus, generate economies of scale (for example, in infrastructure projects), and increase the political weight of the territorial unit in intergovernmental conflicts because it speaks for a larger population.

Should the strategy be successful, claims for more resources acquire greater substance, and dependence on the federal government can be reduced. Municipalities are most likely to increase their scope for action and their political influence on intergovernmental decision-making depends on:

- whether the emergence of 'reproductive interests' (since the mid-1960s) has led to a significant improvement in distribution;
- whether the lead is taken by the large municipalities, i.e. county boroughs (*kreisfreie Städte*) and large county towns (*grosse Kreisstädte*);
- how the municipalities operate and whether they adopt conflictual strategies, mobilising the 'reproductive interests', or adaptive strategies;
- the extent to which the redistribution between the different levels of government reflects changes in the socio-economic structure;
- how far the general trend toward enhancing the role of municipalities is modified by countervailing forces;
- whether the process is structurally significant or whether it is merely the easy stage of a new 'business cycle' in the dynamics of intergovernmental relations.

Analysing the multitude of interdependencies, influences, and interactions between state, municipalities and society is not easy. Obviously, traditional analyses of federalism or intergovernmental relations have to be extended. The inclusion of municipalities permits the analysis of the implementation of federal and state policies, demonstrates the connection between socio-economic interests and intergovernmental structures and, by taking into account different and longer-term approaches to problem-solving, it opens the way for a dynamic perspective on the processes of change. Of particular significance is the evidence that the system has the ability to learn and to adapt, ruling out authoritarian problem-solving (by unilateral redistributions of resources and functions) and encouraging participative and co-operative action. To support this analytical approach to, and extend our understanding of, West German intergovernmental relations, the next section provides a brief account of the evolution of the intergovernmental system in the post-war period.

DEVELOPMENT OF INTERGOVERNMENTAL RELATIONS IN THE FEDERAL REPUBLIC OF GERMANY

The immediate post-war period: occupation and reconstruction 1945–49

After the collapse of the highly centralised National Socialist state, local government was established anew in West Germany. This process was accelerated by the fact that local institutions were the only administrative units still intact and that they resumed their activities immediately after the surrender and took over the essential tasks of public assistance and reconstruction. Furthermore, the occupation powers re-established the institution of local self-government to promote a democratic political culture.[8] The idea of decentralised political processes and institutions found expression in the constitutional regulations of the *Länder* and later in the Basic Law (Article 28, paragraph 2), which guarantees the right of local self-government.

But even in this phase of post-war development, the incorporation of local units into national politics was already under way. The financial resources of the municipalities were completely inadequate. Under the influence of war, the financial system had been transformed into a system of fixed grants with increasing importance attached to revenue sharing. This system was retained almost unchanged after the war.[9] Now the *Länder* assumed the role of the *Reich* as 'allocator' of funds to the local governments. The grants by the *Länder* were of such a volume – amounting to 40 per cent of the total revenue of the municipalities[10] – that the *Land* essentially became the crucial decision-maker over the finances of the localities.

Growth and decentralised policy implementation 1949–66

In the 1950s the Federal Republic of Germany experienced a great economic boom, which increased the prosperity of all social strata. The socio-political developments linked to the boom had important consequences for the municipalities. The gradual development of the welfare state, with its claim to meet comprehensively the needs of citizens, started to reduce the importance and autonomy of local governments. Self-accelerating economic development gradually diminished the service functions which the municipalities had performed in the immediate post-war period. The control of private production was shifted to central institutions.

Pressures for better vertical and horizontal co-ordination of problem-solving and service delivery were also mounting. Among these pressures were the economic imbalance between the *Länder* and municipalities, the inadequate infrastructural supply of certain regions, the overspill problems of the large urban agglomerations and, finally, the tax burden on the *Länder* and municipalities generated by the growth of the welfare state. The call for an 'equalisation of living conditions' throughout the federal territory fostered increased co-operation between the *Länder* and the participation of the federal government in functions performed by the *Länder* and the municipalities.

Given the high degree of social integration in West German society in the first two decades after the Second World War, the conflict between production-oriented and reproductive interests was concealed by the general rise in welfare. Therefore, it was easy to incorporate the local governments into national politics. This incorporation was achieved in three ways:

– the legal regulation and control of local governments;
– the initiation or expansion of the system of delegating tasks (*Auftragsverwaltung*);
– the replacement of block grants by specific (categorical) grants from the federal and *Länder* governments.

The changes reinforced a functional division of labour between local and national governments. Despite their constitutionally guaranteed 'universal competence', the local governments were increasingly confined to dealing with the social consequences of private production and with the supply of local social services.

Although local government increasingly came under the ambit of national decision-makers, there was as yet no direct control of their activities in this period. The dependence of the municipalities on local business tax revenues virtually guaranteed an orientation towards growth; there were no serious deficiencies in the local infrastructure yet, and welfare policies could be managed by the delegation of functions to the local government. As for national housing and economic policies favouring the growth of small firms, the wide latitude given to municipal land-use planning was, at least, not dysfunctional since acute problems could be regulated through the Housing Act (Wohnungsbauförderungsgesetz).

The *Länder* favoured close intergovernmental co-operation as early as the 1950s. Political initiatives were co-ordinated in meetings of the prime ministers (*Ministerpräsidenten*) and of department ministers. Joint demands to the federal government were formulated or, as a rule, jointly discussed with it. Federal-*Länder* meetings attempted to achieve vertical co-ordination between federal and *Länder* governments.[11] These informal meetings were, however, not suitable for binding political programmes. Dissenting *Länder* were able to get round agreements with some ease because the federal government had been given no effective control instruments. It could influence decentralised politics only by granting financial subsidies to lower tiers of governments.

Integration of the municipalities into intergovernmental relations 1966–78

The post-war boom ground to a halt with the recession of 1966/67. Now problems became apparent which previously had been obscured by continuous economic development. Macro-economic management, the harmonisation of social conflicts, the equalisation of horizontal/vertical and social/regional disparities, ecological dilemmas – all required increased governmental intervention which was to bring important changes in the relationship between federal, *Länder* and local governments.

These developments challenged the identity of interests between state institutions and the municipalities. Particularly difficult was the management of the business cycle. The centre's anti-cyclical policies were thwarted by the behaviour of local governments whose revenue structure, despite efforts at reform, was extremely sensitive to economic fluctuations.

About two thirds of overall public investment was handled by local governments, which made them a key influence on business cycle policies. Furthermore, infrastructural shortages appeared: projects were beyond the financial capabilities of the municipalities, thereby encouraging the need for (and the problems of) intergovernmental co-ordination. The dependence of local governments on economic growth became manifest as their problems of fiscal management in a recession became ever more intense.

As a result, the demand grew for supra-local, and partly national, control of decentralised functions. The horizontal co-operation between the *Länder* and between municipalities, which had existed since the early 1950s, was extended and increasingly supplemented by processes of vertical co-ordination and control between the administrative authorities of the different levels. The 1967 Act for the Promotion of Stability and Economic Growth (*Gesetz zur Förderung der Stabilität und des Wachstums der Wirtschaft*) not only introduced the Keynesian macro-economic approach of economic policy-making, but also bound the *Länder* and municipalities to conform with macro-economic targets. The newly established Councils for Economic and Financial Planning (*Konjunkturrat und Finanzplanungsrat*) were required to co-ordinate the budgeting of the territorial units. In addition, federal *Länder* and municipal authorities were incorporated into a hierarchical planning system (covering, *inter alia*, budgeting, infrastructure and physical/regional planning). Although the introduction of legally-binding planning of functions and resources turned out to be impossible, the objectives of federal and *Länder* governments became guidelines for sub-central policies, particularly as the basis for categorical grants.

The 1969 constitutional reform created a new legal basis for intergovernmental relations. Joint planning, decision-making and financing (*Gemeinschaftsaufgaben*) were geared to matching decentralised policies to national targets, arbitrating interregional conflicts, avoiding imbalances between regions, and rationalising the use of resources. Federal grants were made available to help *Länder* and municipalities perform their tasks in housing, urban renewal, urban mass transport and hospital construction. Other reform policies (e.g. the educational reform) were to be carried out on the basis of agreements to be reached between federal and *Länder* governments.

This phase in intergovernmental relations has often been described (and criticised) as a process of centralisation. However, the trend towards joint decision-making (*Politikverflechtung*) between federal, *Länder* and local governments, although restricting local decision-making, did not actually result in functional centralisation. The central government tried only to influence the decision-making of *Länder* and municipalities by means of direct and indirect instruments of guidance and control. The *Länder* and municipalities, for their part, retained extensive rights to participate in the formulation of central policies. Co-ordination and consensus were the staple rule of the intergovernmental game. The federal government could not achieve unilaterally its objectives against the wishes of the *Länder*. In most cases, solidarity between the *Länder* prevented the federal government from playing one off against the other. But, equally, the same held true for relations between *Länder* and municipalities.

In the early 1970s, these systems of co-operation were the means for mobilising resources for specific policy areas, or for protecting these areas against resource transfers to other policy areas. Since consensus between administrative authorities can be achieved only when all benefit, the compromises reached in intergovernmental bargaining were based quite often on transforming problems of redistribution into problems of distribution, thus increasing the demand for resources. Unsurprisingly, there was a sharp rise in general government expenditure from 1970 to 1973. In many cases, the co-operation of federal, *Länder* and local governments led to a 'cartelisation' of administrative authorities. The authorities were thereby able to avoid political conflicts and stabilise the existing distribution of resources.

Joint decision-making was not without its problems. In joint planning, no clear priorities could be established. As a rule, the plans contained imprecise objectives or unrealistic predictions of growth rates. Their effectiveness was, therefore, only marginal. When, in 1973, after the first oil crisis, state resources were tight and growth rates had become lower, joint planning became a substantial problem. Moreover, *Länder* and municipalities protected their jurisdiction from central interference. Any ambitious project and/or attempt at comprehensive integrated planning was blocked or simply ignored in the process of implementation.

Finally, joint decision-making in co-operative federalism strengthened the predominant position of the executive *vis-à-vis* the elected political decision-making bodies. Intergovernmental co-operation expanding into bureaucratic bargaining systems is of low visibility. It led to the blurring of political boundaries and responsibilities. At best, it is partially controlled by parliaments but often there is little or no political accountability. Joint programmes were criticised, therefore, not only as inefficient and inflexible, but also as undemocratic.[12]

THE DYNAMICS OF INTERGOVERNMENTAL RELATIONS SINCE 1978

Problems and perspectives: The revival of 'politics from below'?

Since the end of the 1970s, intergovernmental relations in Germany have been primarily characterised by worsening distributional conflicts between the territorial units due to the growing scarcity of fiscal resources. At the same time, the state has had to cope with increasing problems at all levels of government. Therefore, a number of critics perceive a growing tendency of federal and *Länder* governments to shift expenditure to the lower tiers while redistributing revenue to the centre. Thus, measures to consolidate the budget taken by federal and *Länder* governments were adopted at a time when the municipalities had to expend considerable resources to cope with growing economic and social problems. However, this tendency to 'offload to the periphery' is meeting increased opposition.

As a result of the transformed socio-economic and socio-cultural context, the political importance of local governments, particularly the cities in metropolitan areas, has grown. This change stems from the vertical division of labour between national and local governments whereby the city as bearer of reproductive interests gains in importance with the rise of 'post-material' value systems and is most strongly effected by the economic and ecological crises. Consequently, the politicisation of these issues originated at the local level.[13] The current concern with decentralisation, the 'rediscovery' of local self-government, the politicisation of local politics (particularly in the metropolitan areas) attest to the strength of the transformation.

The demand for a 'revival of politics from below' even occurs in the programmatic statements of the major social groups and of the political parties. Party leaders – conservatives as well as 'greens', representatives of trade associations as well as of citizens' groups – seem to agree that 'politics' has to be revived; that the necessary impetus has to come 'from below'; that the starting point is the 'basics of working and living'; and that these basics must be formulated 'from a decentralised perspective'. This move towards small-scale, regional and even local units probably reflects disaffection with the obvious inefficiencies of centralised policy-making and implementation. Nevertheless, the ubiquity of the call for a political revival anchored at the local level is astonishing. Why has the call fallen on such receptive ears and been taken up in so many quarters?

Part of the answer lies in the changes in the West German party system since the mid-1970s. The 'greens' (or 'alternative' political groupings) succeeded not only in attracting attention to their 'movement' but, with their success in the parliamentary arena, also induced a state of extreme uncertainty in the established parties, especially those parties with a programme and voter base affected directly by the 'new social movements'. The response of the parties is well-known, has been repeatedly commented upon, and warrants only the briefest summary. A phase of attempted stigmatisation and marginalisation of the new political groupings was followed by a thorough-going reassessment by each party of both policy and electoral strategy. These assessments even led to 'heretical' thoughts about forming 'red-green' coalitions. Eventually, *Landtag*-elections in the Saarland and in North Rhine-Westphalia prompted the parties to 'rediscover' their roots. Leading Social Democrats quickly found that the demand for a 'revival of politics from below' was a necessary modification: a new way of formulating policy. Lagging behind, the Christian Democrats and Liberals eventually found that the new politics was an 'enrichment'. However, this 'revival of the political process emerging from the cities and municipalities' was not embraced with the enthusiasm of the Social Democrats.

A second factor is closely linked with the actual performance of the West German federal state. Since the mid-1970s the consequences of tighter public resources, and an increasing number of stalemates in the political and administrative discharge of public functions, have generated more and more criticism. Ungovernability, unproductive interpenetration of the different levels of government, over-regulation and bureaucratisation have been keywords in the debate. The criticisms have called forth demands for policies of consolidation for measures to deregulate and privatise public bodies, and for a return to self-help.

A third approach to the discussion stems from the change in the economic and socio-cultural structure of the Federal Republic. This transformation has faced central and local institutions with the need for significant changes in their routine tasks. Services have to be adapted to meet demographic changes; economic structures have to be modified to cope with the continuing structural change and technological innovations; social security provision has to respond to economic recession; and the natural environment has to be protected. Obviously, economic structural change is a national task, but even here there are pleas to employ decentralised means whenever and wherever possible. There are various references to the 'endogenous potential for development' of a region which has to be 'activated'; to the need for regionalisation, even localisation of policies; and, finally, to the institutionalisation of new private-public partnership schemes in order to increase the 'accuracy' of individual policies and to direct resources to the target groups concerned.

It remains to be seen whether or not the municipalities in the Federal Republic of Germany will be able to prevent the further centralisation of problem management. And it is unclear whether they will succeed in pursuing an aggressive strategy in favour of local interests and in defeating the federal policy of passing on burdens to the municipalities (albeit in the guise of decentralisation). At this juncture, it will be useful to return to the theoretical perspective outlined above in order to assess both the trends in intergovernmental relations and the prospects for a revival of decentralised political and administrative processes.

Trends and processes: an adaptive intergovernmental system

The point of departure must be the fact that processes of policy-making have changed. Although the attractiveness of 'alternative' parties may have passed its apogee, the political issues stirred up by them, and the political initiatives at the decentralised levels, have resulted in a certain re-orientation of the other political parties. The importance of the local level for the political system is taken more seriously than before. Once acknowledged, however, it is as important to note some of the limits to these changes: the discussion of decentralisation might contain irrational expectations; functional and participative demands are partly irreconcilable in a complex intergovernmental system; there are quite a number of substantial problems of guidance and control which are simply not dealt with in the discussion of political decentralisation; the necessity for some degree of intergovernmental consensus and the risk of political stalemates when there are numerous actors in decentralised decision-making processes are problems rarely raised in the discussion.

The decline in federal performance raises questions about 'the potential for revival'. Indisputably, the complexity and entanglement of intergovernmental service delivery systems have led to a number of sub-optimal policy outcomes, and problem-solving seems to be characterised by bureaucratisation. However, it is doubtful whether these tendencies can be reversed without difficulties, and in any case, it is not clear that they are dysfunctional in all respects. Complexity in public services is not an end in itself: it is a reaction to growing demands from an ever more profuse variety of societal groups. Bureaucratic differentiation is a measure of the adaptability and the flexibility of political-administrative institutions. Given the enormous tasks, the degree of differentiation and the interweaving of public and private interests, comprehensive reforms of the structural and procedural components of intergovernmental policy-making seem unlikely. The recommendations submitted by *Länder* commissions on 'debureaucratisation' have covered the simplification of administrative procedures and have made a number of proposals for straightening out regulations, disentangling levels of government and simplifying federal services. However, in order to overcome the inertia of institutionalised interests, such *ad hoc* proposals seem to be inadequate. A continuous and explicit administrative policy is required and at neither the federal nor the *Land* nor the local level is such a policy in the offing.

In so far as the changes in the demographic, economic and social structures first and foremost affect the decentralised territorial units, the call for a 'revival from below' is comprehensible. However, it is not necessarily a logical response. It may be conceded that urban authorities are faced with changed tasks. None the less, it is necessary to pose the question of how these new, or at least modified, tasks can be coped with 'from below'. Certainly, the closeness of the local level to societal problems is invaluable; it is even a prerequisite for an adequate solution. But taking entangled political processes and trying to confine them to a single territorial unit will not guarantee efficient problem-solving or the consideration of wider interests. As long as exchange processes and co-operation between the federal, *Länder*, and local level are a necessary prerequisite for intergovernmental policy-making, relatively simplistic ideas of reform remain unworkable if not dangerous. They underestimate the dynamic interaction between economic and social developments and political-administrative reactions and they ignore the demand for joint decision-making.

This brief survey of current developments in the relations between federal, state, and local governments[14] has shown that the call for a 'revival of politics from below' is to some extent surprising. The analysis identified some good reasons for emphasising decentralised politics and administration: closeness to the problems, orientation at target groups, and regional and sectoral 'accuracy' of policies are but a few. It is also not disputed that the local level is appropriate for mobilisation and innovation, for countering the separation of state and citizen and for reminding the national level of its responsibility by referring problems to the federal or *Länder* government. But are the revivalists' expectations realistic? Do the complex processes of policy-making, of management and of control in the welfare state permit a simplified model of decision-making? There might be the danger of falling into the trap of romantic ideals of comprehensive self-government and co-determination rather than taking into account the institutional and procedural complexity of 'post-industrial' politics and policies.

A series of case-studies on the roles of urban agglomerations in the process of policy-formulation in the federal state of West Germany[15] concluded that structural reforms of the intergovernmental system are both improbable and unlikely to succeed. It is true that the relations between central territorial units (federal and *Länder* governments) and local governments are characterised by increasing antagonism. These distributional conflicts have their roots not only in institutional self-interests, but also in the increasingly different socio-economic interests attached to each level of government (i.e. the conflicts between 'reproductive' and 'production-oriented' interests have intensified under the conditions of economic crises). This situation has given rise to challenges to the existing structural patterns and routines of intergovernmental policy-making but, at the same time, the high costs of transforming institutional structures require conflict-reducing decision rules.

As noted in our analytical approach, intergovernmental processes of redistribution can be brought about either by politicisation outside the intergovernmental system or by modes of co-operative-participative bargaining. In times of tighter finances and worsening conflicts of distribution, the willingness of individual actors to articulate their interests outside the intergovernmental system and to find coalition partners increases. Despite severe conflicts over distribution, however, intergovernmental bargaining systems remain stable and continue to work. The parliamentary arena is used either for exercising vetos or for highlighting the financial problems of all local authorities. Party influence on intergovernmental distribution processes is reduced when the affected policy is long-standing, complex and problem-solving is highly structured. Only when there is close party identification with certain groups of municipalities and when distributional processes can be redefined so as to be exploited in election campaigns (because they directly affect voters) are the political parties involved. They act essentially as a 'transmitter' rather than as additional problem-solvers. They react to a situation that is already politicised rather than seeking to politicise a situation.

Politicisation of intergovernmental relations is most likely in defence of existing positions. Local governments can increase their influence if they open intergovernmental processes of distribution to external political pressure. Certainly there is evidence that they are able to maintain the status quo and effectively defend themselves against initiatives from the federal or the *Länder* governments. Indeed, this ability virtually compels the central authorities to co-operate and bargain with them.

These processes of co-operation and bargaining are highly sensitive to changes in 'political paradigms'; that is, in those collective values, experiences and attitudes that reflect changes in economic, social and cultural conditions. Certainly, bargaining processes which are quasi-institutionalised, permanent and have been reduced to a system of actors of considerable continuity are sensitive to such changes (for instance, in the area of financial equalisation [*Finanzausgleich*] and possibly for the so-called 'permanent administrative reform'). It is not change in structures but in processes (that is, in the modes of co-operation and of joint decision-making in the intergovernmental system) which are of importance in understanding the intergovernmental dynamics of the West German system. By these means, intergovernmental relations adapt to changed socio-economic conditions and to changes in demands in the functions carried out by political institutions.

The call for a 'revival of politics from below' not only expresses hopes and expectations for greater efficiency, transparency and legitimacy in traditional politics by employing decentralised political processes, it also recognises that the relations between state and society are changing again. Over the last few years it has become apparent that the steering, guidance and control devices used to cope with societal problems in a regulative and authoritative way, relying on skeleton planning, orders and restraints, have become less and less efficient. Since policy-making is characterised by a large number of new actors representing special interests and by changed forms of political articulation, policy implementation must be sensitive to the range of horizontal and vertical actors and to the respective target groups. Quantitative policies are no longer adequate and have to be complemented by qualitative policies; the 'hardware' of politics has been partly replaced by its 'software'. Motivation, communication, acceptance and dialogue are keywords of the current debate: a participative-co-operative model of politics aims for the increasing scarce resource of 'consensus'.

In this context, relations between city and state, between central and sub-central territorial units have also obviously changed. Superior–subordinate relations and the hierarchic allocation of functions according to the hierarchical structure of the state have become more difficult. New demands for communication and co-operation have changed the routines of intergovernmental policy-making. And, despite its limited scope of action, the local level is gaining ground and is becoming – as an authority of implementation, co-ordination and integration – an increasingly important part of national policy-making and problem-solving:

- with respect to *structures*, due to the increased importance of urban agglomerations. Economic, socio-cultural and structural change has enhanced the importance of urban areas as centres of political, administrative, economic and cultural development and innovation, particularly in times of tighter resources;
- with respect to *processes*, as a result of the central position of decentralised territorial units in the implementation of federal and *Land* policies, and of current trends in the decentralisation and regionalisation of some policy areas – a consequence of the deficient routines of central problem-solving and/or attempts to shift burdens to lower levels;
- with respect to *substance*, as massive social problems at the local level have intensified the municipalities' role in political integration, which is of especial significance given the current changes in the process of political articulation.

Any further evaluation of the intergovernmental system of the Federal Republic of Germany will not be able to ignore these factors.

CONCLUSIONS

This analysis suggests findings of interest not only for research into federalism but also for the practice of intergovernmental relations. First of all, it is clear that, in the case of the Federal Republic of Germany, we have to assume a dynamic intergovernmental process which goes beyond existing constitutional norms and which cannot be adequately grasped by the traditional static analysis of federalism. Dynamics in this sense means analysing developments in the federal system not only in their structures, but also in their processes and policies. It means extending the analysis of interactions between the different levels of government to include informal processes of participation, the capacities of institutions to learn and to adapt to changed situations, and, finally, to variables fashioned by history and culture such as 'policy-styles'. Such a perspective avoids the unproductive 'labelling' of different intergovernmental systems and focuses on the description, analysis and assessment of actual developments in federal states. Describing an intergovernmental system as 'unitary' or as 'federal' provides little useful information. It says nothing about its actual performance or its capacities for problem-solving and for integration. The same point can be made about characterising a decision structure as 'territorial' or 'sectoral' or a resource structure as 'revenue-division' or as 'revenue-sharing'.

The study of central-local government relations reveals that changes in the intergovernmental system of the Federal Republic were the result of the central government's power to define the situation and to formulate corresponding policies, but they were *not* an expression of quasi-'imperialistic' dominance. That reforms were initiated 'from above' is due to the share in the responsibility of the federal level for the production of some public goods at the subordinate level. It also has the structural advantage of being able to define problems and to set the framework for problem-solving in a quasi-monopolistic position. The lower levels, on the other hand, have to bear high costs for reaching the consensus and the cartelisation required for the accommodation of a great number of heterogeneous interests. Only a limited part of the structural power of the central government can, therefore, be deployed in German intergovernmental politics. The concern to reduce tension in the intergovernmental system is accorded higher priority than the (short-term) advantages gained by the unilateral exercise of power. The high price of consensus and the existence of powerful structural veto-positions explain why the costs of transaction in reform processes serve as the most relevant indicator of 'good' solutions. In other words, the political process moves towards solutions as the several actors learn and adapt their aspirations. These solutions may be described as genuine reforms but are characterised by the effort to keep the costs of transaction as low as possible.

It is not surprising, therefore, that the structures of the intergovernmental system in the Federal Republic of Germany tend to produce conservative or, rather, conserving patterns of politics which are stabilised by factors such as institutionalised learning and political paradigms. Although 'politics' play an important role in the process of defining the disparities requiring action and in the process of reaching consensus, their impact is reduced by the pre-existing limited scope of action: existing patterns of distribution of power and interests prevail; the spectrum of alternatives is determined by the power of the central state to define the framework of institutional and organisational constraints, norms of action and accepted definitions and interpretations of problems. 'New' solutions are, therefore, variations of standard solutions. The actual outcome is the result of bargaining and learning, in the course of which the actors lower their level of aspiration and strive to minimise the costs of transactions.

NOTES

1. Speyer (Schriftenreihe der Hochschule für Verwaltungswissenschaften Speyer) *Politikverflechtung zwischen Bund, Ländern und Gemeinden* (Berlin: Duncker and Humblot, 1975); Fritz W. Scharpf, Bernd Reissert and Fritz Schnabel, *Politikverflechtung. Theorie und Empirie des kooperativen Föderalismus in der Bundesrepublik Deutschland* (Kronbert: Scriptor, 1976); Joachim Jens Hesse (ed.), *Politikverflechtung im föderativen Staat* (Baden-Baden: Nomos 1978).
2. Anna Kraus, *Zentrale und dezentrale Tendenzen im Föderalismus* (Göttingen: Vandenhoeck und Ruprecht, 1983), p. 199.
3. Bundesministerium der Finanzen, *Die Finanzbeziehungen zwischen Bund, Ländern und Gemeinden aus finanzverfassungsrechtlicher und finanzwirtschaftlicher Sicht* (Bonn: BMF, 1982). Finanzbericht, *herausgegeben vom Bundesministerium für Finanzen* (Bonn: Heger, 1985), pp. 113–14 [figures refer to 1981].

4. Scharpf *et al.*, *Politikverflechtung*.
5. F. W. Scharpf, *The Joint-Decision Trap: Lessons from German Federalism and European Integration* (Berlin: Wissenschaftszentrum, Discussion Paper IM/LMP 85–1, 1981).
6. Dietrich Fürst, J. J. Hesse and Hartmut Richter, *Stadt und Staat. Verdichtungsräume im Prozess der föderalstaatlichen Problemverarbeitung* (Baden-Baden: Nomos, 1984); Arthur Benz and J. J. Hesse, *Zur Dynamik intergouvernementaler Beziehungen – Ein Analyseansatz* (Speyer, mimeo 1986), p. 30.
7. D. Fürst and J. J. Hesse, 'Zentralisierung oder Dezentralisierung politischer Problemverarbeitung? Zur Krise der Politikverflechtung in der Bundesrepublik', in Hesse (ed.), *Politikverflechtung im föderativen Staat*, p. 191; Fürst and Hesse, 'Die Funktion von Verdichtungsräumen im Prozess der Politikverflechtung', in *Ballung und öffentliche Finanzen* (Hannover: Schroedel-Forschungs und Sitzungsberichte der Akademie für Raumforschung und Landesplanung, Bd. 134, 1980), p. 165; Fürst, Hesse and Richter, *Stadt und Staat*.
8. Theodor Eschenburg, *Jahre der Besatzung 1945–1949* (Stuttgart/Wiesbaden: DVA/ Brockhaus – Geschichte der Bundesrepublik Deutschland, Band 1, 1983), p. 245.
9. Herbert Sattler, 'Gemeindliche Finanzverfassung. Bedeutung gemeindlicher Finanzhoheit für die Selbstverwaltung', in *Handbuch der kommunalen Wissenschaft und Praxis, Band III* (Berlin: Springer, 1959), pp. 1–30.
10. Herbert Bohmann, *Das Gemeindefinanzsystem* (Köln: Kohlhammer; Neue Schriften des Deutschen Städtetages, H. 2, 1967), p. 52.
11. Renate Kunze, *Kooperativer Föderalismus in der Bundesrepublik* (Stuttgart: G. W. Fischer, 1968).
12. Scharpf *et al.*, *Politikverflechtung; F. W. Scharpf, B. Reissert and F. Schnabel (eds.), Politikverflechtung II. Kritik und Berichte aus der Praxis* (Kronberg: Athenäum, 1977); Hesse (ed.), *Politikverflechtung im föderativen Staat;* Frido Wagener, 'Zur Zukunft des Föderalismus und der kommunalen Selbstverwaltung', in *Der Landkreis 1981* (Köln: Kohlhammer, 1981), p. 105. Arthur Benz, *Föderalismus als dynamisches System* (Opladen: Westdeutscher Verlag, 1985), p. 30.
13. Fürst, Hesse and Richter, *Stadt und Staat*.
14. For details on different policy areas, see J. J. Hesse (ed.), *Erneuerung der Politik 'von unten'? Stadtpolitik und kommunale Selbstverwaltung in Umbruch* (Opladen: Westdeutscher Verlag, 1986).
15. Fürst and Hesse, 'Die Funktion von Verdichtungsräumen im Prozess der Politikverflechtung', p. 165; Fürst, Hesse and Richter, *Stadt und Staat*.

Italy – Territorial Politics in the Post-War Years: The Case of Regional Reform

Robert Leonardi, Raffaella Y. Nanetti and Robert D. Putnam

The history of territorial politics in post-war Italy has been dominated by the gradual – though not by any means linear – movement towards a restructuring of the country's institutions. During the last 40 years, the Italian state system has moved away from a centralised to a 'mixed' model – that is, neither a centralised nor a federal form of intergovernmental relations. In practice, we have seen a significant regionalisation of the institutional policy process and political and interest group activity. This contribution analyses the *political* as well as *institutional* debate which has accompanied the transition and describes the significant turning points in the process by which the regional reform has changed. It focuses on: the structure of the policy-making process; the nature of intergovernmental relations; political and interest group processes; the array of 'actors' and the institutional and political strategies adopted by the centre to limit 'policy slippage' and inverse the tendency towards 'policy inertia' of policy-makers.

Any discussion of territorial politics in Italy must, by necessity, deal in large part with the politics of regional decentralisation because it has been the focus of 'the reform of the state' debate since the demise of Fascism. When the Constituent Assembly met in 1946 to draft a new Constitution for the Italian Republic one of the primary goals was to replace the highly centralised regime inherited from the Fascist and Liberal periods with a decentralised regional state system. In addition, all of the other proposals to restructure local government from the area-wide planning districts to neighbourhoods and metropolitan government have mostly emerged as by-products of regional reform.

However, the importance and implications of the regional decentralisation drive in Italy do not have the same connotations as territorial cleavage patterns in other countries. Italy's territorial cleavages are not similar to the cultural, linguistic or ethnic differences found in other countries in Western Europe, such as Spain, Belgium, or Switzerland.[1]

Consequently, the concepts and terms used in this study need to be clarified to counter the ever-present danger of different meanings attendant on any discussion of territorial politics in diverse political settings (see Rhodes and Wright, above). As is illustrated in Figure 1, the nature of the Italian state structure and relations among various levels of government have fluctuated over time and can best be described as being governed by a hodgepodge of administrative traditions and approaches that have appeared during the last 50 years. For example, the legislation that regulates the activities of the provinces and communes was passed during the Fascist period (in 1934) and pre-dates the 1947 Italian Republican Constitution. The Constitutional Court came into existence nine years after the Constitution was ratified, government policy-making was in full operation, and four regional governments had already been created.[2] Before 1970, the intermediate level of government identified in Figure 1 as 'regional' had been implemented in only five out of the 20 regions in the country. And certain organs that exist at the national level for the purpose of interacting with sub-national governments – such as the Prime Minister's Regional Office, the Ministry for the Regions, and the Interparliamentary Committee on the Regions – were created in 1970.

FIGURE 1 THE INSTITUTIONS OF ITALIAN INTERGOVERNMENTAL RELATIONS

Level of Government	Function Executive	Legislative	Oversight and Judicial Review
National a = 1 b = 1947	Ministry of Interior; Ministry for the Regions; Prime Minister	Interparliamentary Committee on Regions; Other Committees	Prime Minister's Regional Office; Constitutional Court; Court of Accounts
Regional a = 20 b = 1947/1977	President and *Giunta*	Council and Committees	Regional Control Commission; Court of Accounts; Regional Commissioner
Provincial a = 95 b = 1934	President and *Giunta*	Council and Committees	Prefect; Provincial Control Committees
Communal a = 8,085 b = 1934	President and *Giunta*	Council and Committees	Prefect; Sectional Control Committees

Key: a = number of governmental units located at this level.
      b = data of legislation that regulates activities of level of government.
Note: There are also other territorial subdivisions – such as mountain communities, health service districts, school districts, transportation districts, neighbourhoods, etc. – existing at the sub-provincial and sub-communal levels.

After 1972 the provincial prefects lost their predominant oversight role with regard to provincial and communal government activities to provincial committees and sections of the Regional Control Commissions whose members are appointed by various organs of the regional government (i.e. regional president and council), and oversight of regional legislative activities is conducted in the 15 ordinary regions and Val d'Aosta by the Regional Commissioner appointed by the central government (in many cases he is the prefect of the province where the regional capital is located). Instead, in four of the 'special' regions (Sicily, Sardinia, Trentino-Alto Adige, and Friuli-Venezia Giulia) that function is performed by the regional Court of Accounts.

Since 1971, other sub-provincial and sub-communal territorial subdivisions have been created by national and regional legislation to manage the implementation of sectoral policies, such as the national health care system, education, and the development of mountain areas. All these developments suggest that despite the emphasis on deductive legalistic thinking on the part of Italian scholars of intergovernmental relations[3] the reform and institutionalisation of new sub-national organs and relations with national government have developed in a quite inductive and piecemeal fashion. The reality of Italian intergovernmental relations reflect the frequent recourse on the part of political leaders to political compromise and the adoption of a gradualist approach in implementing their programme. As expressed by Sergio Bartole with regard to the regions, 'the regional reform was not thought out at one point in time, according to a guiding idea; it was instead realized — and is still being realized — over time through a long and difficult process of negotiation'.[4]

A case in point is the way that intergovernmental relations have been mediated in the 1970s through the creation of a whole series of 'mixed' committees bringing together national and sub-national government representatives to manage sectoral programmes. Before 1970, the primary links between the national and local levels of government were the administrative/oversight relationship embodied by the provincial prefects and the Minister of the Interior and the role of the individual parliamentarian in the political sphere. In the 1970s both these linkage systems declined in importance. The role of the prefects/Minister of the Interior has been replaced by the Prime Minister's Regional Office to which the Regional Commissioner (who represents the state at the regional level) reports. The prefects have been forced to maintain their provincial responsibilities, but their legislative oversight function has been substantially taken over by the Regional Control Commission.

In the political sphere there has been a considerable expansion of alternative political linkages through the creation of the Interparliamentary Commission on the Regions, the Ministry for the Regions, and, after 1981, the creation of a Conference of Regional Presidents that meets directly with the Prime Minister and selected ministers. The communal and provincial governments have not been able to institutionalise their contacts with the national level through formal political organs, but they have, none the less, been effective in lobbying the centre through their local government interest groups represented by ANCI and UPI.[5] Thus, intergovernmental relations in Italy have expanded substantially during the last 15 years, and more is expected in the near future.[6] But all these innovations point to the serious decline in the importance of the role of administrative controls and the emergence of other arenas (political, financial, legal, party, and interest group organisations) in the management of intergovernmental relations.

It is with this changing reality in mind that we must clearly define the concepts that will be used in discussing the implementation of the regional government reform in Italy after 1970 and the nature of intergovernmental relations in this new setting. First of all, 'decentralisation' will be used to describe the overall 'ideal' or 'goal' to be pursued in the restructuring of the Italian state. The implementation of the decentralisation goal requires the fulfilment of numerous objectives and institutional change, such as new political institutions, new relations among the political forces, the integration of new economic and social actors, and the affirmation of a new set of political and institutional values in intergovernmental affairs. The concept of 'devolution' of power to sub-national government is one of the specific objectives that is central to the process of decentralisation. Devolution refers to the change in the distribution of political and administrative decision-making power in favour of the sub-national political units.

In this context, the two other terms — 'transfer' and 'delegation' of powers — are associated with specific implementation strategies adopted to achieve the devolution of power and the final objective of local government reform. More specifically, it is proper to use the term transfer in referring to the policy areas outlined in Article 117 of the Italian Constitution which are primarily the responsibility of the regions; the delegation of power is the method used to interpret, in light of modern developments, the 'intentions' rather than the specific 'dictates' of the Constitution and allocate to local and regional governments new areas of responsibility previously handled exclusively by the national government. For analytical purposes, it is useful to distinguish two aspects of the process of decentralisation: devolution of authority from the centre to the regions and the effective utilisation of that authority by the new empowered units. Devolution and utilisation are, of course, ultimately related, both in the obvious sense that what is not devolved cannot be utilised, and in the somewhat less obvious sense that the pace of devolution is likely to depend on the effectiveness with which existing grants of authority are exploited. Nevertheless, this distinction is important theoretically, since devolution and utilisation are likely to be affected by different variables. In particular, uniformly delegated powers might be utilised with widely varying degrees of success in different regions or localities. Such, in fact, has been the experience of the Italian regions.

The process of devolution of decision-making powers means the *increased* (rather than absolute) autonomy in the policy process of sub-national units of government. In general, autonomy refers to the capacity for discretionary decision-making, unhindered by external control. In formal terms, intergovernmental autonomy is affected by factors such as the legal and constitutional framework, the administrative framework (administrative controls, delegated functions, personnel patterns, and so on), and financial provisions for supporting locally determined policies. But these factors — laws, rules, and money — are significant, but not absolute, constraints on local government autonomy, and they are, in turn, resources in the endless, intricate bargaining that lies at the core of intergovernmental relations everywhere.

TERRITORIAL POLITICS IN THE IMMEDIATE POST-WAR PERIOD

Although regionalist sentiment was strong during the initial period of Italian unification, it was only after the Second World War that regionalism became dominant and received active support from a broad spectrum of political forces. The implementation of regional decentralisation was seen as a means of democratising the state apparatus and ensuring access to the policy process by diverse socio-economic and political groups in society. As expressed by Michele Monaco, the goal of regionalising the Italian state emerged from the conviction that 'a well-functioning' and 'democratic' state had, by necessity, to have a 'decentralised rather than a centralised structure'.[7] Thus, from the beginning, the movement for decentralisation had two goals that were inextricably linked: the rationalisation and reorganisation of the national bureaucracy and the devolution of powers to the regions and other units of local government.

However, during the Constituent Assembly debate, it became clear that the major parties disagreed on how strong the proposed regional governments should be and that this difference in attitudes would affect the pace at which the regional decentralisation was inacted. Once the Constitution was drafted, the dual goals of decentralisation − regional and bureaucratic reform − were not implemented as expected; they fell victim to the partisan political conflict that erupted in the post-1947 period and became institutionalised in the subsequent hegemony exercised by the Christian Democratic Party (DC) over the Italian political system. The history of the decentralisation/regionalisation struggle in Italy can be divided into five periods that have been characterised by different phases of territorial politics:

- 1946–60: the period of 'centrist' regionalism that saw the creation of four of the five special regions and the passage of the 1953 Scelba law that imposed a stringent control from the centre on any future regional reform;
- 1960–70: the period of 'centre-left' regionalism that completed the implementation of the special regions, began the legislative process that brought into existence the ordinary regions in 1970, and created regional planning bodies;
- 1970–72: the 'constituent' phase of the regional reform and the initial transfer of power and personnel to the nascent regions;
- 1973–77: the mobilisation of the regionalist front that led to a qualitative expansion of regional powers and empowerment of local government; and
- After 1978: the 'management' phase of the regional reform when the regions had to assume the responsibility for a wide range of policies.

Our analysis focuses on the developments and ramifications of the regional reform on government policies and strategies in the years following the first nation-wide regional elections in 1970. After the regional elections the nature of the devolution game changed dramatically. What was before only a hope for reform has become a reality; what had remained on the level of abstract theory about regional performance has been filled with empirical evidence; and what were expectations about the model for centre-periphery relations has been tested by the course of events.

1970–72: THE 'CONSTITUENT PHASE' AND THE FIRST TRANSFER OF REGIONAL POWERS.

The regional elections of 7 June 1970 represented a historic watershed in the development of Italian political institutions, and it was a major step in the full implementation of the provisions of the 1947 Constitution and the goal of restructuring and modernising the Italian state. The Communist Party (PCI) saw the regions as the first step on the road towards a thorough reform of the state.[8] The creation of a national system of regional governments was for the PCI an essential element in the continued democratisation of the state and the advance towards a more egalitarian society. As Pietro Ingrao declared on the eve of the 1970 electoral campaign:

> We are interested in the regions as political assemblies that are engaged in the general reform of the state and society and through this reform we want to diffuse political power and make it more pertinent to the needs of the masses.[9]

The emphasis of the 1970 PCI campaign was on the realisation of 'a new way of governing' and where possible the creation of 'open regional governments' (*giunte aperte*). The first slogan was the expression of the need that the Communists felt to open up the governing process at the regional level to forces outside of the traditional institutions and governing parties – e.g. excluded interest groups, voluntary associations, and individual citizens. The Communists felt that this approach to regional decision-making would lead to a change in the quality and quantity of demands transmitted by the citizenry to the regions and other parts of the political system – i.e. at the national as well as local levels of government.

Sentiment toward the regions was also very strong among the Socialists (PSI). In the 1970 elections the Socialist leaders had a dual objective: to show their party base that despite the increased conservatism of the DC they had not abandoned their commitment to reform and co-operation where possible with the Communists in setting up leftist regional *giunte*; and to emphasise the essential role played by the party in the constitution of the governing coalition of the left, as well as the centre–left variety, in promoting effective decision-making and in emphasising the 'centrality' of the PSI in local and national government alliances. The combined effect of these two goals was to commit the PSI not to enter any coalition where its votes in the council were not essential to the formation of a governing majority.

The Christian Democratic view of the regions was decidedly mixed. In general, the party conducted its 1970 campaign on the basis that the regional reform was a necessary element in the implementation of the Constitution, but in some regions the level of enthusiasm varied according to the factions that were in the ascendency at the local level. In parts of the north where the *Base* and *Forze Nuove* were in control of the regional party organisation, enthusiasm for the reform was quite high. In these areas, such as Lombardy and Emilia-Romagna, the DC was committed to the cause of state reform and the democratisation of the Italian political system, especially with regard to the promotion of increased citizen participation (which had become one of the principal objectives of the regions in 1970), and the DC was not at all reluctant to open a dialogue with the PCI in areas where the Communists were the dominant party. For these DC members, the creation of the regions was seen as the institutional response to the student and workers' movements of the late 1960s and the need to break down the old ideological/institutional barriers for the purpose of initiating a dialogue between the major representatives of Italy's Catholic and Marxist forces. In the words of one DC regional councillor:

> 1970 came after the 1968 student movement and 1969 worker movement, and the regions were also conceived as institutions ideally suited and capable of increasing citizen participation, intensifying the political debate, and permitting a breakthrough in the congested political situation

at the national level by involving a wider range of political and social forces in a new attempt to re-think the structure of Italian society: a new exercise of the will and imagination to correct and mollify the congested and blocked national political situation.[10]

The 1970 regional election results showed the PCI improving its position by one percentage point over its 1968 showing, while the DC slipped back 1.2 per cent in the two years from the last parliamentary elections. The PCI made gains in five north-central regions (Tuscany, Emilia-Romagna, Liguria, Lombardy and Veneto), equalled its 1969 vote in one (Umbria) and fell slightly in the two remaining north-central regions (i.e. Piedmont and Marche) and all of the seven southern regions. Christian Democratic Party fortunes also showed variation from one part of the country to the other: it showed gains in three regions (Molise, Campania and Umbria), no change in one (Piedmont), and losses in all the rest. The biggest winners in the election were the Social Democrats (7.0 per cent), Italian Social Movement (6.0 per cent), and Republicans (2.9 per cent) while the parties that were most penalised by the results were the Liberals (-1.1 per cent), PSIUP (-1.1 per cent), and Socialists who saw their total drop from 13.4 to 10.4 per cent.

The distribution of council seats allowed the Socialists to implement their governing coalition strategy in 12 of the 15 regions. Of the three leftist *giunte* formed in 1970, the PSI stayed out of only one (Emilia-Romagna) where the PCI and the Proletarian Unity (PSIUP) had a one vote majority. In DC-dominated areas the Socialists did not enter into two *giunte* (Veneto and Molise) where the Christian Democrats had absolute majorities. In all the remaining regions, centre–left coalition governments were constituted.[11]

Once the governing coalitions were formed, regions adopted different procedures in allocating roles within the regional institutions. In Emilia-Romagna, as part of the operationalisation of the 'open *giunta*' concept, important positions in the council were allocated to parties which were not members of the governing majority. Silvano Armaroli of the PSI was elected president of the council while Christian Democrats and Republicans were given the presidencies of council standing committees. The approach adopted by the DC one-party *giunta* in Veneto was much different: it allocated to itself all the important council posts. In the other regions there was a sharing of important posts among the parties constituting the majority – for example, in Puglia where the DC assumed the presidency of the *giunta* (Trisorio-Liuzzi) and the PSI, with Finocchiaro the presidency of the council, and in Tuscany where Lagorio (PSI) became *giunta* president and Gabbuggiani (PCI) council president – though there continued to be constant problems among the *giunta* partners over the relative allocation of council and *giunta* posts.

Despite this jostling for advantage within the regional governments, the first two years of the regional experience were characterised by a substantial amount of 'consensus building' in the management of regional activities. The most important problem that had to be faced during the second half of 1970 was the formulation of the regional statutes, and this was done with considerable unanimity. According to Galgano and Pellicani,[12] the only party that consistently opposed the acceptance of the regional statutes as formulated by the 15 ordinary regions was the MSI. And even it joined the passage of the regional statute in Basilicata (though abstaining on the preamble).

The 1970 regional elections transformed regional decentralisation into an open competition with the introduction of new 'actors' — that is, from the regions themselves to the new Ministry for the Regions and the Interparliamentary Commission for the Regions — and with the presence of an independent arbiter in the form of the Constitutional Court that could intervene in an authoritative manner in the disputes between the regions and the centre.

The tensions between the two levels of government were not long in emerging when the regions, in formulating their statutes, clearly refused to be bound by the 'minimalist' approach codified in Law 281/1970. Instead, the regions took as their model the broader approach enunciated in the 1947 Constitution. A number of statutes stated that the regions conceived themselves as an instrument for the devolution of state power, the promotion of popular participation in the formulation of public policy, and the initiation of a rationalisation of socio-economic policies through planning and innovative policy formulations. The regions did not define themselves as mere instruments for the administration of predetermined national policies. In fact, regional decentralisation was seen as part of the process for finding solutions to national economic and social problems: the development of the south, the elimination of socio-economic disequilibria, and the integration into national life of those forces (e.g. emigrants, youth) that had long borne the cost of distorted national priorities.

Soon after the statutes were accepted by Parliament, a number of regions went on the attack by calling into question Article 17 of Law 281/1970 that gave the state 'the function of directing and co-ordinating the activities of the regions that pertain to exigencies of a unitary nature, also in reference to the objectives of national economic planning and commitments derived from international agreements'. The law also provided for the allocation of this power of initiative and co-ordination to various branches of the national government: 'the exercise of the function outlined above can be delegated from time to time by the council of ministers to the interministerial committee for economic planning (CIPE) for the purpose of determining the operating criteria for its areas of responsibility or to the Prime Minister and the interested minister when it has to do with particular problems'. The regions argued that they, and not national organs/agencies, had been empowered by Articles 115 and 117 of the Constitution to determine policy in a number of specific areas. Therefore, Article 17 of 281/1970 was, in their opinion, unconstitutional. In 1971 the Constitutional Court did not support the regional position on the matter, but the significance of the challenge was that the outcome of the devolution game was no longer determined exclusively by the bureaucracy or selected elements of the executive branch of national government.

The culminating act in the state's attempt to minimise the change in the policy game that was suggested by the creation of the regions was the passage of the *decreti delegati* of 1972 that fully operationalised the goals of the minimalist strategy. According to Barbera and Bassanini, the 1972 decrees cast the regions as:

... a sort of superprovince, a peripheral terminal for central administration, empowered to manage, above all, separate and fragmented functions, in particular those services and those activities which the ministries had preferred to 'abandon' without having decentralized choices and decisions of any political significance.[13]

Despite the restrictive nature of the 1972 decrees, the 'constituent' phase did provide signs of a change in tactics of the minimalist forces in the attempt to contain the regional reform. The combined impact of 281/1970 and the 1972 decrees reaffirmed the role of the state — that is, the importance of the administrative linkages — in controlling the activities of the regions, but they were a different set of administrative linkages (e.g. Regional Offices at the Council of Ministers and the Regional Commissioners) than the one (Ministry of the Interior and prefectoral system) traditionally used to exercise central control. The shift in emphasis served to undermine the previous power and status of the provincial prefects. This trend was illustrated by the comments of a southern prefect who, in a personal interview, observed that the region had had a subtle, subversive influence on the perception of power by the people. Opening his arms toward the piazza in front of the prefecture, he said: 'Before the creation of the region this square would have been jammed with people lining up to ask for favours or advice on day-to-day problems. Now look at it. It is empty!'

1972–77: THE MOBILISATION OF THE REGIONALIST FRONT

On 4 July 1972, the second Giulio Andreotti government (1972–73) presented to the Senate a routine request to extend the deadline for its decrees on the reorganisation of the state bureaucracy as envisaged by Law 775/1970. Article 5 of the law had not been implemented due to the lack of two decrees on the restructuring and reduction of the central ministries. Both the Communists and Socialists criticised the government for requesting an extension after the original deadline had already passed and for not presenting any clear indication of how it was going to proceed on the matter. The government's position was further weakened when its request for an extension found the unexpected support of the radical right. It now became easy for the regionalists to draw clear ideological distinctions between the rightists' support for the government's minimalist strategy and the left's attempt to block that strategy.

On 17 May 1973, the Senate voted by 149 to 140 against the government's proposal. At this point, the tide of thinking had slowly begun to turn in favour of the regions and the positions championed by the left. The impact of the change in mood was immediately evident in the contents of the Senate's version of the alternative legislation which proposed to cut and reorganise the central bureaucracy with the aim of strengthening the regional governments. For their part, the regions were now in a position to voice their demands for, and expectations of, a more complete decentralisation of powers. They were able to insert some of their own ideas and wishes into the Senate version of the bill.

After the issuing of the 1972 decrees, a few regional leaders, cultural figures, jurists and politicians had tried to create a 'regionalist front' that could effectively press national decision-makers for a more advanced form of regional decentralisation. The regional governments of Lombardy, Emilia-Romagna and Tuscany were particularly active in this period. The former was in a favourable position to lead the regionalist front due to its dynamic socio-economic fabric, quality of leadership and availability of resources. Since the end of the war, Emilia-Romagna had performed a vanguard role for the left in experimenting with new policies at the local level. A parallel move in the direction of greater regional autonomy was also taken by the new actors created by the regional reform — i.e. the Ministry and the Interparliamentary Commission for the regions.[14]

In September 1973 the Interparliamentary Commission began a series of hearings to obtain the views of the region on the future of the reform. The strong regionalist tone of the hearings had a clear impact on the first draft of the bill presented to the Senate which constituted a major shift in content of the delegation of authority that the government was asking.

Pressure began to mount at the beginning of 1974 within the First Commission that was assigned the task of redrafting the *legge delega* as well as within the ranks of the governing parties to bring about a full transfer of powers to the regions. The task was effectively assumed by the First Commission of the Senate, which produced a draft acceptable to Parliament and the government and received widespread support when it was voted on 26 July 1974. The government did not assume a direct role in the discussions of the contents of the new bill and gave the Senate Commission a free reign. Oversight of the government's position was delegated to the Minister for Relations with Parliament (Silvio Gava in the fourth Rumor government and Luigi Gui in the fifth Rumor government).[15] The 1974 Senate bill had similar characteristics to the law that was eventually to become the watershed legislation, Law 382 of 1975, which moved beyond the initial, rather restricted position of bureaucratic reform towards the goal of an organic transfer of powers to the regions. The initial Senate version of the bill proposed to treat the special regions in the same manner as the regular regions. But at the explicit request of the special regions, they were excluded from subsequent reformulations of the bill in the Chamber of Deputies. The argument used by the five special regions was that if they were placed on the same plane with the other 15 they would lose the special status allocated to them by the Constitution.[16]

It was in the course of the debate in the Chamber that the separation of the two issues of regional decentralisation and bureaucratic reform took place once again. This time, in contrast to the 1970 legislation, the government was not able to allocate the task of paring down the central bureaucracy's power to that same bureaucracy. The Chamber insisted on maintaining parliamentary oversight of the process by specifying that the government had to get the approval of the Interparliamentary Commission before it could turn its proposals into law. It was felt that this innovation would keep the subsequent formulation of the executive decrees on a highly political plane, where regionalist forces could bring their maximum strength to bear and prevent the decrees from being undermined by the bureaucracy. Control of the arena for decision-making was correctly seen by the regionalists as a key to the political dynamics of devolution.

At the regional level, the discussion in the Chamber took place at the time when the first terms of the legislatures were ending. In this second round of elections, the left as a whole (Communists +5.5 per cent and Socialists +1.6 per cent) made gains, while the Christian Democratic Party lost 2.5 per cent in comparison with its 1970 performance. Even though there were many different reasons for this change in voting patterns, the respective positions of the winners and losers on regionalism seems to provide added weight to the interpretation that, in part, the electoral result was a popular expression of support for a more rapid and thorough devolution of power to the regions. As a direct consequence of this change in political climate, on 22 July 1975 Parliament passed Law 382 that codified all the sentiments for a more thorough regional reform that had been building up since 1973. Law 382 was passed with the support of all the parties, except for the PLI and MSI.

Within the regional institutions the impact of the 1975 elections served to change the way of conceiving regional alliances and policy formulation — that is to say, they instituted what virtually amounted to a second 'constituent' phase. Once again, the emphasis was a consensus-building and an attenuation of the distinction between the majority and minority forces in the council. The PSI no longer felt bound to the existing governmental formula at the national level and opened negotiations to enter into governmental alliances with the Communists in a number of regions — for instance, Emilia-Romagna, Piedmont, Lazio and Liguria — where the two parties had not previously shared power. The DC, in turn, responded by opening the distribution of council posts and the formulation of regional policies to both the PSI in areas where it had enjoyed absolute majorities (Molise and Veneto) and to the PCI in traditional centre−left regions — that is, putting into practice a DC version of the *'giunte aperte'* advocated previously by the PCI. The effect of these political changes at the regional level had an impact on the national debate on the regions, which was immediately evident in the qualitative leap represented by the contents of Law 382.

M.S. Giannini has described 382/1975 as a new form of law: 'It is a law that interprets and integrates the Constitution' rather than an ordinary formulation of national policies affecting specific sectors.[17] The uniqueness of the legislation is in its prescribed 'methodology' for interpreting the meaning and intent of the relevant constitutional provisions,[18] in spelling out the process to be followed in formulating the legislative decrees, and providing for a diversity of consultations in coming up with the final version of the decrees.

As required by one of the constitutional transitions/dispositions, Law 382 delegated to the government the responsibility for completing the devolution of power, and stipulated the following criteria:

- all powers attributed by the Constitution to the regions that were currently being exercised by the ministries and/or national agencies were to be transferred in an organic rather than piecemeal fashion;
- the remaining administrative functions that had not been passed down in 1972 were to be decentralised;

- all state expenditures that directly or indirectly were attributable to the regions were to be eliminated as a block and not through piecemeal legislation;
- all agencies and functions in areas of regional responsibility that bridged more than one region were to be transferred;
- the regions were to have full administrative and legislative powers within the limitations of national legislation and EEC directives.[19]

In contrast to the 1970 laws, 382/1975 provided for a number of reviews – one on the part of the regions and two on the part of the Interparliamentary Commission – before the decrees could be transformed into law. In this manner, the regions would not be left to the mercy of the state bureaucracy. The law required that the initial draft of the decrees be prepared by the government; that the regions present their opinions on the draft within 60 days; and that the Interparliamentary Commission make its observations within the subsequent 60 days, whereupon a second government draft of the decrees would be drawn up which, in turn, would be reviewed by the Interparliamentary Commission, before the final definitive decrees were presented by 25 July 1977. To these six steps were added two others. One step, before the formulation of the first government draft, consisted of a report by a blue ribbon commission, chaired by Giannini, which was given the task of providing expert opinion of how the decrees should be formulated – thereby introducing another substantial limitation on the government's ability to limit the impact of the regional decrees.[20] The second step was the creation of a working group, composed of three politicians [the three spokesmen for the DC (Signorello), PSI (Aniasi), and PCI (Cossutta) responsible for local government affairs] and three jurists (D'Onofrio, Bassanini and Barbera), which negotiated the final version of the decrees.[21]

The results of the 1976 elections reinforced the trends of 1975, and one of the most important results at the national government level was the slow transformation of the Communists from the chief opposition party to a member of the governing coalition. Almost immediately after the election PCI representatives were allocated posts as presidents of standing committees in Parliament, and Pietro Ingrao was elected president of the Chamber of Deputies. PCI support became indispensable for government stability. The 'historic compromise' strategy launched in 1973 by the party Secretary, Enrico Berlinguer, and designed to forge a political alliance among the Communists, Socialists, and Christian Democrats for the purpose of carrying out basic reforms had received a major boost. The effects of this relationship were immediately felt in the subsequent discussion of the decrees implementing Law 382.

The first draft of the decrees was presented to the regions of 18 February 1977. After three weeks of studying the proposals the regions informed Parliament that the decrees were unacceptable in either quantitative or qualitative terms and did not conform with the objectives of Law 382: 'What is necessary, therefore, is more organic rather than sectoral powers, sustained by an adequate source of funding in order to launch a serious planning process in the areas of regional responsibility'.[22]

The first governmental version of the decrees represented an attempt to abort the maximalist approach to the regional reform that had characterised Law 382 and the recommendations of the Giannini Commission. One Christian Democratic regional councillor observed that the proposed decrees 'do not adhere to either the letter or the spirit of Law 382'.[23] Giannini asserted that the origin of the government's draft remained a mystery: no one was willing to acknowledge paternity.[24] The limited orientation of the first version was evident in the government's forecast of having to transfer 93 billion lire to the regions as a result of the devolution of powers. The Giannini Commission had foreseen a transfer of 120 billion, and the regions in their subsequent evaluation came up with a figure of 200 billion. As a result, several members of the Interparliamentary Commission agreed with the view expressed by the regions and undertook to rewrite a good part of the decrees,[25] and the Commission's draft provided four-fifths of the version that was finally promulgated.

Added weight was given to the work of the Commission by the agreement of 25 June 1977 in which all of the parties supporting the Andreotti government bound themselves to accept the version of the decrees supplied by the Interparliamentary Commission. This accord was not completely respected as the government sat down on 8 July to consider its response to the Commission's new draft. The Minister of Agriculture, Giovanni Marcora, was particularly incensed by the massive transfer of powers in the agricultural sector. In his opinion, the regions were in no way prepared to manage such a large set of new powers effectively.

The text that returned to the Interparliamentary Commission modified some of the Commission's proposals, especially in the area of social assistance where the Church had expressed major concern about provisions restricting public support of religious organisations. On this matter, the Socialists insisted that local governments be given exclusive control of public funds destined for social assistance programmes. A compromise was reached in the Commission, but the Socialists were not completely satisfied. The PSI accused the Communists of sacrificing the completeness of the regional reform to its historical compromise strategy of not alienating the DC. Despite the dispute between the Communists and Socialists over the inclusion of religious organisations in the administration of certain social services, the Commission's second review of the decrees found it in substantial agreement with the changes made by the government. On 24 June President Leone signed the decrees.

After 30 years of waiting, a substantial devolution of regional powers was achieved based on a political accord among the three major parties. Guido Fanti, PCI president of the Interparliamentary Commission on the Regions, noted the difficulties of trying to maintain harmony with the Commission so that the objectives of the reform would not be lost in the sometimes bitter exchanges between Commission members.

> How would it have been possible to achieve a similar result if we had not tenaciously and persistently negotiated with the DC, leaving aside the simple propagandistic trick of hardening our position on everything by running the risk of destroying such an important reform? The dilemma

of the Commission was, therefore, to insist on the search for an agreement that would force the Government to accept our conclusions and enact the important parts of the reform or to privilege some particular aspects, even at the risk of compromising the stability of the political accord or the Government. The unitary nature of the effect was on the whole maintained.[26]

The decrees, DPR 616, went into effect on 1 January 1978. In putting some of the final touches, the government attempted to postpone the decentralisation of a few national agencies until there was an explicit law covering the subject, but the overall impact of the decrees was not diminished. In financial terms the decrees gave the regions control of approximately 25 per cent of the entire national budget with some estimates running as high as one-third; 20,000 offices were abolished or transferred to the regions, and 15 general directorates and a large number of divisions within the ministries were abolished. Another innovation that was inserted in the decrees was the stipulation that deadlines were set for the passage of a whole series of reforms, from reform of the university system to public health, local government, and local finance.

AFTER 1978: THE 'MANAGEMENT' PHASE

The 1977 decrees were unique in a number of ways: not only in the way they were formulated but also in the way that they defined regional powers. They implemented a creative interpretation of Article 117 of the Constitution and reflected modern conceptions of socio-economic policy-making by grouping together organic sectors and thereby recognising the interrelationship of social needs and institutional functions. The 1977 decrees identified three main policy areas of sectors for which the region had primary responsibility: territorial planning, social services and economic development. The weak point of this revolutionary redefinition of regional responsibilities was that there was little echo of this change at the national level – that is, with the implementation of 616 there was no parallel radical change in the central government's structure or mode of behaviour. There was no paring down of the central administrative apparatus, change in the legislative process to take into account the regions' primary responsibility in a number of policy areas, or provisions for the consultation of the regions during the deliberations of national decision-making centres – such as the Council of Ministers or Parliament. In addition, no change was made in the structure of other organs vital for the correct functioning of the newly decentralised apparatus – for example, the Constitutional Court and the Court of Accounts.[27]

One of the first examples of the implications for the regional reform of this missing link was the state's lack of response to Article 11 of the 1977 decrees stipulating the regions' role in the formulation of the national plan and financial policies. In 1978, the Pandolfi Plan practically ignored the existence of the regions and Article 34 of its own financial legislation (468/1978) stipulating consultations with the regions in formulating the national budget. Even worse was the government's budgetary proposal for 1979 which foresaw, once again, a cut in allocations to the regions and an increase to those ministries such as transport and public works that were severely affected by the regional devolution of power. In addition, the government postponed the crucial reform of local government bodies, thereby undermining that part of the innovative thrust of the 1977 decrees which called on local governments to increase their capacity for programmatic intervention and forcing the regions to continue emphasising its 'administrative' approach to policy-making rather than being able to make a qualitative change in the direction outlined by the 616 decrees.

For its part, Parliament also failed to demonstrate a full understanding on the new distribution of powers in its formulation of national legislation. The most obvious areas where Parliament showed little understanding of the regions' areas of responsibility were its forays into agriculture, industrial reconversion and transport. The constitutional obstacle was most often bypassed through the creation of state-regional commissions for the co-determination of policies. By the early 1980s, there were 99 such commissions. Nor was there any attempt to change parliamentary procedures so that standing committee deliberations could solicit regional views when policies involving the regions were under consideration or required a 'consultative' opinion on the part of the Interparliamentary Commission on the Regions. The regional reform, therefore, still had not resolved the problem of providing an organic institutional link between the regions and Parliament.

At the regional level the post-616 period saw intense activity to implement the goals of the decrees. But it was also the period in which there was a deepening of the economic crisis and the collapse of the coalition of national solidarity when the Communists walked out of the governing coalition in 1979. The shift of the national coalition to an expanded five-party centre−left (fifth Andreotti and first Cossiga governments) sparked a series of crises at the regional level which had the effect of bringing to an end the *'giunte aperte'* arrangement that had been operating since 1975; increasing the tensions between the Communists and their Socialists allies; and re-establishing the need to create uniform centre−left *giunte* where the Socialists had abandoned leftist alliances. Until 1980 there was no immediate Socialist response to the change in the system of alliances at the national level although, in Campagna, the Socialists left the *giunta* after the collapse of the programmatic accord with the PCI. However, the situation in the country crystallised in 1980 after the third round of regional elections.

In those regional elections the Communist Party lost half the gains it had made in 1975. The most worrying losses were registered in the south and some areas of the north. The Communist vote remained stable in the centre at the levels reached in 1975. The DC recuperated well in those areas (e.g., Lazio) where it had showed considerable weakness in 1975−76. However, it did not fully reestablish its image of power, self-confidence, or votes to the level generated in the 1970 elections (36.9 per cent in 1980 in comparison with 37.8 per cent in 1970). The real 'political winner' of the election was the Socialist Party which, despite increasing its vote *vis-à-vis* 1975 by only 0.6 per cent, gained an enormous amount of self-confidence and clout. Thus, the PSI was able to negotiate forcefully with its allies advantageous political arrangements. Of particular importance was the emergence of strong party regional leadership in Marche, Calabria, Lazio, Liguria and Piedmont. In two of the regions previously ruled by a PSI−PCI coalition, the increased friction between the two leftist parties was resolved by the creation of PCI one-party, *monocolore* governments: Emilia-Romagna in 1980 and in Tuscany in 1983. In general, the third legislature was to be characterised by an increased amount of conflict between the PCI and the parties of the national centre−left coalition.

In the period between the end of the 1970s and the beginning of the 1980s the regions had to face the consequences of the economic crisis which increasingly exercised pressure to reduce public spending at a time when the regions were being asked to assume new responsibilities. In 1978 the Ministers for the Treasury and for the Budget imposed a reduction in regional and local spending. The regions reacted by insisting that such action not only constituted interference in their constitutional powers but was also an attempt to undermine their efforts to gain control over regional and local spending by giving priority to their activities.

In response to these attempts, the regions concentrated on three objectives in redefining their relations with the national government. First, they sought to 'penetrate' the centre of national decision-making (the Council of Ministers) institutionalising direct political dialogue with the national representatives, placing the stress on the political nature of the dialogue and abandoning the alternative of administrative relations that had proved so conflict-ridden and for so long had dominated centre−regional relations. Second, they emphasised the reform of local government. This reform would allow the regions to resolve the problem that they had no 'intermediate' level of government for the implementation of a planning approach to policy-making at the regional level. Finally, they insisted that local government finance be restructured. In this way, sub-national levels of government would know in advance the size of their budgets and, therefore, be in a position to finalise investment programmes.

Andreotti had been the first Prime Minister to favour an institutional mechanism for permanent dialogue between the state and the regions. In 1980 the Interparliamentary Commission suggested the creation of a 'state-region conference' that would bring together on a regular basis the 20 regional presidents, the Prime Minister and the Ministers for the Treasury, Regions, and Budget. It also suggested that the institutional site for the conference be the Council of Ministers. For the regionalists, the institutionalisation of a permanent presence of the regions at the centre became an absolute necessity in stopping attempts by the state through the power of the purse to weaken the regional reform. Franco Bassanini observed in 1983 that:

> We are passing through a period of increasingly limited resources in relation to the services and public policies that society demands. If today we were to conserve a strict separation between the regions and the central organs of the state, the regions may have the formal appearance of autonomy but, in fact, would find themselves only capable of administering the cuts in spending imposed by the state. Therefore, either we guarantee the regions' and other local governments' participation in the process where resources are distributed or else the regional reform will be emptied of its content.[28]

Another reason for creating the Conference of Regional Presidents was provided by the need to rationalise and control the proliferation of state–region consultative organs. On 12 October 1983 the Conference of Regional Presidents was instituted within the Council of Ministers. This step was the first move towards adopting a model of state-regional relations that emphasised co-participation in decision-making and interdependence in the administration of policies at the two levels of government. This model is qualitatively different from the one that emphasised the separation of sectoral responsibilities and which was dominant during the first ten years of the regional reform.

The effects of the recent economic crisis and the decrease in funds available to regional and local government have accentuated the need to reform their finances if the responsibility of remaining within the confines of locally-generated resources and budgetary projections is to be a realistic obligation. In 1972–73 the state undertook to reform local government finances in the light of the changes wrought by regional reform. But instead of transferring more responsibilities to local government bodies in the collection and management of tax revenues, the thrust of the legislation moved in the opposite direction: it centralised to the national level all taxing powers and made local governments almost completely dependent on 'derived' or 'transferred' revenues. This state of affairs has made it increasingly difficult for the regions and all other types of local governments to prepare realistic three-year spending programmes and investment budgets as called for by Law 335/1976 or manage their yearly budgets. A case in point was provided by the delay in announcing the state budget in 1982[29] which literally blocked any budgetary initiative at the regional level. The Minister of the Regions at that time, Aldo Aniasi, argued that the redefinition of financial arrangements between the state and the regions was the key means of resolving the impasse.[30] However, the course of the debate and actions taken by the centre on local and regional finances well illustrates that the struggle over decentralisation and regional autonomy has not been settled and that its focus can shift from one moment to another to different arenas. The devolution of powers from the state to sub-national governments must be seen as a dynamic process rather than a specific institutional arrangement that remains stable over time.

At present, the themes of particular interest within the debate on the reform of sub-national government are those concerning the new role for the province, the institutionalisation through *ad hoc* mechanisms of relations between the regions and local government, and, while maintaining the centrality of communal government, promoting forms of association among them. In 1985 a local government reform bill was passed by the Senate, but it may be some time before the new law finally clears all of the hurdles necessary for passage and implementation.

CONCLUSIONS

Looking back over the last 40 years, it seems clear that the regional reform has not succeeded in achieving all the multifaceted goals of decentralisation implicit in the Constitution. But it has succeeded in devolving a number of organic powers to the sub-national level and substantially changing the nature of intergovernmental relations. This objective was achieved through three strategies: the increase in the number of relevant actors, the change in the nature and, therefore, rules of the process, and finally in changing the relationships among the principal actors.

With regard to the first strategy, it has been shown that after 1970 there was an expansion of actors at both the national as well as regional level. At the centre, new actors were created – e.g. the Interparliamentary Commission and the Ministry for the Regions – while existing actors – e.g. the Constitutional Court – were given important roles in the structuring of relations between the centre and the periphery. And, as a consequence, other national actors – such as the Ministry of the Interior and the provincial prefects – lost a considerable part of their power in determining local policy choices. In response to the decline in importance and efficacy of administrative/bureaucratic controls, the focus of activities among national leaders to control sub-national institutions has shifted to the manipulation of political (political party leaders), legislative (Parliament), and financial (budgetary process) resources whose allocation is still determined at the centre. Thus, the centre still has a predominant role to play in the determination of intergovernmental relations, but the nature of the relationship has changed substantially in the post-1970 period through the entrance into the game of a whole series of new actors.

Considerable changes also took place at the regional level. The most important changes was the creation of a regional political system that brought into existence a new regional political class, policy process, and parallel aggregations of interest groups. Before 1970 with the exception of the 'special' regions, the regional level of politics was practically non-existent. The creation of a potentially powerful set of regional political institutions has nurtured the growth of a whole series of parallel organs in the political, economic, and social spheres of collective action, and the emergence of these new levels of activity has had a profound impact in changing the relations and structures of actors outside the formal governmental institutions.

With regard to the second strategy, and especially after 1977, the rules have been changed so that devolution is no longer a simple confrontation between the state and the regions, but has become a multifaceted bargaining arrangement involving the state, the regions, other local governmental bodies, and the new array of socio-economic and political groups that have been reorganised at the regional level. The expansion of alternative linkage systems available to regional actors for influencing national policy also means that the nature of the rules governing interactions are under constant flux and redefinition.

Finally, the change in the nature of the process means that, with regard to the third strategy, the relations among the three institutional players and new party and interest group players are much more complex. Thus, the conflict arising between levels, such as between the state and the regions and between the regions and local governments, can be considered physiological rather than pathological in nature – that is, they are integral parts of the new rules and game. It follows that the resolution of these conflicts is no longer purely juridical or administrative in nature, but rather has become an issue for political bargaining. To phrase it another way, relationships among different levels of government in Italy have assumed the same characteristics that for a long time have shaped the relations among similar levels of government in federal states in other Western democracies: juridical relations have given way to political ones in determining the outcomes of conflicts among different levels of government.

NOTES

1. A comparative study of élite cleavage patterns conducted in 1976 by G. Loewenberg (eds.), *The Role of Parliament in the Management of Social Conflict* (Durham, N.C.: Duke University Press, forthcoming), covering parliamentarians in Belgium, Switzerland and Italy revealed that the Italian élites were separated by one cleavage (ideology), while élites in the other two systems were divided along a number of different cleavages, e.g. cultural-linguistic, ethnic and religious, in addition to ideological.
2. In fact, the Sicilian Regional Statute was adopted in 1946 before the Republican Constitution was ratified in 1947.
3. B. Dente, 'Centre-Local Relations in Italy: The Impact of the Legal and Political Structures' in Y. Mény and V. Wright (eds.), *Centre-Periphery Relations in Western Europe* (London: Allen & Unwin 1985), pp. 125–48.
4. S. Bartole, 'Il caso italiano', *Le regioni*, Vol. 3 (May–June) 1984, pp. 411–29, p. 415.
5. The regionalisation of interest groups and the generation of diverse regional economic dynamics are some of the more interesting consequences of regional decentralisation during the last 15 years.
6. There is now a proposal to link all the regional councils with Parliament in a formal manner because, until now, the relations between the centre and the region have been channelled through the executive branches of regional and national government.
7. M. Monaco, *La regione* (Rome: Edizioni Cinque Lune 1957), p. 11.
8. G. Berti, 'La riforma dello stato' in L. Graziano and S. Tarrow (eds.), *La crisi italiana 1979*, pp. 447–92.
9. P. Ingrao, *Masse e potere* (Rome: Editori Ruiniti 1977), p. 311.
10. L. Melandri, *Regione Emilia-Romagna*, June 1981, pp. 40–2.
11. In Basilicata for the first two years there was a one-party DC minority *giunta* that depended on fluctuating support from the right (PLI and MSI) as well as the centre−left (PSI and PSDI). In 1972 an organic centre−left coalition was formed.
12. F. Galgano and F. Pellacani, *Statuti regionali comparati* (Bologna: Zanichelli 1972), Chapter IV.
13. A. Barbera and F. Bassanini (eds.), *I nuovi poteri delle regioni e degli enti locali* (Bologna: Il Mulino, 1978), p. 22.
14. It should be added that the trade union movement and private economic interests were also members of this regionalist front.
15. The Minister for the Regions at that time was Mario Toros, a strong advocate of the maximalist approach to regionalism.
16. At that time, it was not at all clear that Parliament would pass a bill that would give the ordinary regions more power and resources than had been granted initially to the special regions.
17. M. S. Giannini, 'Del lavare la testa all'asino' in Barbera and Bassanini (eds.), pp. 7–18.
18. Article 1 of the Law states: 'The identification of the policy areas must be carried out according to organic sectors. This must be done not on the basis of ministerial responsibilities or those of local organs of the state or other public institutions but according to objective criteria that are deductible from the full range that these sectors have and to the close relationship that exists between complementary structural functions. Those transfers of power should be complete and fulfil the purpose of ensuring a systematic and programmatic management of the territorial and social responsibilities that constitutionally belong to the regions' (A. Barbera, *Governo locale e riforma dello stato* (Rome, Editori Riuniti, 1978), p. 128.
19. E. Modica and R. Triva, *Dizionario delle autonomie locali* (Rome: Editori Riuniti, 1974), pp. 815–21.

20. The Giannini Commission was created by Toros and Morlino in 1974–75 to evaluate what would be the best procedure to follow in devolving powers to the regions. The Commission completed its work in December 1976 and presented its results at a conference in Milan on 28–29 January 1977. *Regioni Italiane, Il completamento dell 'ordinamento regionale per il rinnovamento e la riforma delle istituzioni*, Atti del convegno sull 'attuazione della legge 382/1975 (Milan 28–29 January) (Bologna, Il Mulino, 1977).
21. The time allowed for this procedure was one year, but before the year was up, a governmental crisis in January 1976 brought on new parliamentary elections in the summer of 1976. Thus, formulation of the regional decrees was postponed until 25 July 1977 with the passage of Law 894/1976.
22. Camera dei Deputati, *L'attuazione della '382'* (Rome: Camera dei Deputati-Servizio per i rapporti con le regioni 1977), p. 23.
23. Camera dei Deputati, *L'attuazione della '382'*, p. 4.
24. Giannini, p. 15.
25. Camera dei Deputati, *L'attuazione della '382'*, pp. 10–11.
26. Ibid.
27. This state of affairs brings to mind the image used by Giannini, 'He who washes the head of a donkey wastes both his time and soap': that is to say, one can never fully achieve the transfer of powers to the regions because there is always some part missing, and in most cases that missing part is the reform of the bureaucracy.
28. F. Fassanini, 'Bisogna partecipare alle scelte economiche', in *L'Italia delle regioni*, 26 (October–November), 1983, pp. 6–7.
29. The information was finally provided in April rather than in December as had been promised and required by law.
30. A. Aniasi, *Rapporto 1982 sullo stato delle autonomie* (Rome: Istituto poligrafico dello Stato, 1982), pp. 18–19.

The Netherlands: A Decentralised Unitary State in a Welfare Society

Theo A.J. Toonen

The basic structure of the Dutch intergovernmental system is currently characterised as a decentralised unitary state. The Netherlands are no exception to the rule that intergovernmental relations (IGR) traditionally have been the domain of scholars in the field of public and administrative law. One cannot blame the latter for their emphasis on the importance of subnational structures and processes. However, the predominance of a public law orientation has limited the analytical insights available. This contribution explores the topic from an interorganisation or multi-actor perspective.

The organisation of this overview is as follows. In order to understand some main features of contemporary IGR it is necessary to go back into history, since the basic structures which constitute the intergovernmental system of the Netherlands were founded in the middle of the last century. Despite efforts to 'reorganise', 'rationalise', and 'modernise' the system in the 1960s and 1970s, the constitutional and legal structure laid down in the 1850s still applies. Section 2 provides a description of the main actors, the legal framework and some developments which are important in understanding Dutch IGR. Section 3 is an overview of developments that took place in the subnational and intergovernmental system with the growth of the welfare state; developments generally presented as an ever increasing process of centralisation. In section 4 there is a critical assessment of this centralisation thesis from a multi-actor perspective. By contrasting these two perspectives an interpretation of IGR in a 'decentralised unitary state' is provided. Section 5 concludes with a short summary.[1]

IGR IN THE NETHERLANDS: ACTORS, FRAMEWORK AND DEVELOPMENTS

The constitutional and legal framework of Dutch IGR was established in the middle of the last century. Despite changes and adjustments, the main structure laid down in the Constitution of 1848 (and subsequent organic legislation), the Provincial Government Act of 1850 and the Municipal Government Act of 1851 still applies.

Actors in IGR

The Dutch intergovernmental system, known as the system of Home Administration, consists of three layers of government: a municipal, a provincial and a national level of government.[2] At present there are 12 provinces and about 740 municipalities. The key relationship in the system is between central and municipal government, given that, in the Netherlands, 'local government' is virtually a synonym for 'municipal government'.

Municipalities

Municipalities are general purpose territorial governments.[3] According to the Municipal Government Act, the municipal council is the highest authority within the municipal government structure. The size of municipal councils ranges from seven to 45 council members depending on the number of inhabitants in the municipality. Council members are chosen by direct elections for a period of four years.

In terms of power and daily government and administration, the core of municipal government is the municipal executive. The executive body consists of a Burgomaster (mayor) and aldermen. The aldermen have to be members of the municipal council. They are elected by the council every four years after the local elections. A municipality with a population of 20,000 or less has two aldermen. Municipalities with 20,000 – 100,000 inhabitants have three or four aldermen. In municipalities with a population over 100,000 the municipal council may decide to have four, five or six aldermen.

The Burgomaster is appointed by the national government for a period of six years; a second term of office is possible. The Burgomaster is the chairman of the municipal council (without a vote) and the chairman of the municipal executive (with a casting vote in case of a tie). In addition, the Burgomaster has formal duties and powers in his own right. He is head of the police, with a specific responsibility for the maintenance of public order, and he is responsible for the implementation of various national statutes.

The distribution of power within the municipal executive largely depends on personal factors and city size. In smaller cities the power is likely to lie with the Burgomaster and the highest civil servant (the Town Clerk). In larger cities these actors are still important, but power is more likely to be vested in the sectoral, policy divisions of the municipal bureaucracy, with the aldermen as the main political figures.

Provinces

The Dutch provinces vary considerably in size, population and economic structure. Their boundaries are largely a product of historical circumstances. Recently, a twelfth province was installed in the reclaimed land area of the former Zuiderzee. As at the municipal level, the provincial council is formally the highest authority. The size of provincial councils varies from 47 to 83 members. The provincial executive consists of persons elected by the provincial council from its own membership for four years. The executive body and the provincial council are chaired by a Queen's Commissioner appointed by the national government.

Provincial government has its own jurisdiction, tasks and policies. However, its position in IGR is best typified as that of intermediary *cum* co-ordinator between national government and municipalities. Provinces play an important role in, for example, physical and regional economic planning and in policy areas such as housing and environmental affairs.

The Provincial Government Act entrusts the provincial executives with the task of supervising the municipalities and Water Control Boards. Within this legal framework a variety of provincial – municipal relationships has developed, ranging from a rather distant, authoritarian approach in some provinces to an interactive co-operative approach in others.[4] Provincial budgets are relatively small, as the bulk of subnational tasks and policies are carried out by municipalities.

National Government

In IGR the important actors at the national level are the ministries. There are few formal and direct linkages between the different representative and executive bodies at the subnational level and Parliament (the Second Chamber). To talk about central–local relationships in the Netherlands is to talk about the interactions of the executive side of government.

At present, there are 13 ministries. The number is not fixed and can be changed in the process of government formation. Each ministry is headed by a minister who is its political head as well as a member of the Council of Ministers or Cabinet. The council is chaired by the Prime Minister who is the political leader of the Cabinet. Formally, members of the Council of Ministers have equal power. The Council of Ministers and its supporting administrative structure of co-ordinating committees are supposed to be the overarching and integrative institutions within the national government bureaucracy. Actually, and despite the principle of collegial government and administration, this structure has not resulted in a tight, hierarchically structured bureaucracy. The several ministries and policy divisions within them play a relatively independent role. The structure is sometimes referred to as the 'kingdom of the thirteen disunited ministries'. A lack of co-ordination, fragmentation and overlap has been the recurrent subject of criticism by several authors and governmental committees.[5] For the purpose of this analysis it is sufficient to note that local and provincial governments are dealing with a fragmented national government structure.

LEGAL FRAMEWORK

The constitutional and legal position of municipalities within the decentralised unitary state is characterised by autonomy and co-government.

Autonomy

In the Constitution, the domain of municipal government is loosely formulated. Localities are responsible for the government and administration of their 'own household' which is 'left to' municipalities (and for that matter, provinces). Municipalities have a right to take the initiative in conducting their own affairs. The domain and scope of autonomous local government activity are not directly determined by national or provincial government. They are, however, indirectly influenced. The Dutch system of public law is based on a principle of hierarchy. In the case of conflicting legislation, national and provincial acts rank above a municipal government act. In effect, this means that, formally, the bounds of the municipal 'household' can be determined only negatively, like everything which is not being regulated by higher authority.

In exercising their right to initiative, municipalities are subjected to provincial supervision: municipal plans, proposals and budgets need the approval of provinces. In turn, provinces often require the approval of the national government when, in exercising their supervisory powers, they wish to turn down municipal proposals. Supervision is not a command relationship: it involves the application of 'negative' sanctions such as suspension, nullification or the withholding of approval to municipal decisions. It enables provinces to tell municipalities what they should not do, not what they should do.

The decentralised unitary state bestows higher authorities with 'blocking power'. The characteristic of being one's own master is the capacity to take the initiative. In the original conception of the Dutch unitary state this power was vested with municipalities.

Co-government

A consequence of the imprecise demarcation of the municipal domain is that there is no fixed or objective standard to determine what belongs to the municipal 'household' and what does not. This gives municipalities opportunities to open up and enter new policy-areas. Under the Dutch system of public law there are, however, no constitutional or other legal guarantees for municipal government. Any and every municipal initiative can be regulated by national legislation.

In drafting the Constitution the reformers looked for other ways of securing the position of provincial and municipal governments. One such device was the principle of (provincial and) municipal self-administration. Thus, it was argued that national legislation did not imply implementation by a bureaucracy of national civil servants. The constitutional and subsequent organic legislation should create opportunities for provinces and municipalities to implement national legislation. Most important, implementation was not interpreted as a neutral, mechanistic activity. It was seen as a value-laden activity and as a source of power. Consequently, by giving the municipalities the power to implement national legislation, the position of local governments within the state structure would be strengthened. And by reinforcing their position, the possible emergence of a nationwide bureaucracy could also be averted.

The principle of self-administration has created a lot of heated debate and corresponding confusion over the years, and its significance has been eroded gradually. Currently the arrangements for municipalities to implement national legislation are referred to as *medebewind*, translated here, for want of a better term, as 'co-government'. This change in terminology reflects the erosion of the original conception of self-administration. In current debates about IGR, *medebewind* refers to the provisions for central intervention in local affairs.

The initial conception of Dutch IGR can be summarised as a *compound structure*. Municipalities have a right to initiate in conducting their own affairs. In exercising this municipal autonomy, they depend on the approval (non-resistance) of provincial supervisors. In many instances, national legislation depends on other governmental and non-governmental bodies, including municipalities, for its implementation. In short, the municipal field of activity is characterised by the rules of autonomy and co-government.

DEVELOPMENTS: TOWARDS A WELFARE SOCIETY

This compound structure has not been immune from legal and socio-economic change. To understand contemporary Dutch IGR the gradual expansion in both the quantity and the intensity of relations and contacts between the different layers of government have to be described.

Organic legislation

There has been relatively little change to the basic framework of organic legislation. Although the Local Government Act of 1851 has been amended many times these changes have been minor. However, there have been important developments *within* the basic legal framework of central-local finances. In 1865 local excises were abolished. Instead, municipalities were given a general grant which amounted to 80 per cent of the personal taxes collected by the national government in the municipalities' area. At first, this grant varied with tax-yields. In 1885 the system was changed to a fixed sum, and in 1897 the system was changed into a payment per inhabitant. The most drastic change in the central-local financial relationships took place in 1929 when the *Gemeente-fonds* (Municipality-fund) was introduced. This fund is a form of revenue-sharing in which a certain percentage (currently around 11 per cent) of national taxes is allocated to a general fund and then distributed between the municipalities, utilising an objective formula. The Council of Municipal Finances, an advisory body composed of representatives from both levels of government, plays an important role in administering the Municipality-fund.

The transfers from the *Gemeente-fonds* are called general grants, and are allocated to municipalities without strings. Apart from some detailed changes, in, for example, the allocation-mechanisms, the Municipality-fund remains the main instrument in central-local finances. In consequence, local taxes form only a small part of local revenue, amounting to some 6–12 per cent of total municipal income.[6]

Specific legislation

Besides organic legislation, legislation in the different policy-sectors has affected the position of local government. Since the beginning of the century, this specific legislation has been at the heart not only of the expansion in co-government, but also of its changing character. Originally, co-government referred to rather mechanistic tasks, such as organising (national) elections or administering the recruitment procedures for the military draft. Gradually, national government ceased to set the rules and regulations governing implementation. Instead, it directs municipalities to regulate certain areas rather than carrying out the regulations enacted by higher authorities.

Both quantitative and qualitative changes have reversed the situation prevailing in the 1850s: the majority of municipal activities fall under the heading of co-government, whereas autonomy characterises only a very minor proportion.

Socio-economic developments

The expansion of co-government can be interpreted as a response to the interweaving of activities and levels of government under the pressure of socio-economic development.

In the Netherlands, the Industrial Revolution arrived relatively late compared with other West European countries. Its impact was felt mainly after the 1870s. Thereafter population growth, industrialisation, urbanisation, growing economies of scale and increasing mobility and communication quickly changed the landscape of the intergovernmental system. Despite the prevalence of *laissez-faire* ideas at the time, government actively participated in several of these processes. By the end of the nineteenth century local governments were actively engaged in gas production (for streetlights) and other public utilities: e.g. waterworks, telephone, public transport and electricity.

Besides abolishing barriers for a free economic development, national government played an important role in improving the infrastructure (railways and canals) – improving, for example, the accessibility of the Rotterdam (1872) and Amsterdam (1876) harbours. Among other things, it was also actively engaged in agriculture and shipping, especially in times of economic crises.

The turn of the century saw the first social security legislation: the beginnings of the welfare society. At the beginning of this century government as a whole spent about 10 per cent of the national income. Thereafter the public sector expanded in a way comparable with most other West European societies. Comparatively, the Netherlands belongs to that category of countries in which government expenditure in the twentieth century grew spectacularly or, depending on one's point of view, alarmingly.[7] In the 1980s about 70 per cent of the GNP is being spent by government in all its guises. As Table 1 demonstrates, the growth in government expenditure was especially great in the period after 1955.

TABLE 1
PUBLIC SECTOR EXPENDITURES IN THE NETHERLANDS
1950–84

Year	Governmental Expenditure (%GNP)	Transfer payments (%GNP)	Total (%GNP)
1950	16.8	14.7	31.5
1955	19.0	15.4	34.4
1960	18.3	17.6	35.9
1965	21.1	20.3	41.4
1970	22.1	26.1	48.2
1975	23.8	36.1	59.9
1980	22.6	40.7	63.3
1984	21.0	48.9	69.9

* Government expenditure: direct expenditure in terms of wages, salaries, investments and government consumption.

Transfer payments: mainly payments for social security and interest public debt.

Source: Ministry of Finance, 1984.

The growth of the welfare society peaked between 1960 and the mid-1970s. The bulk of the growth was caused by increased transfer payments. Direct expenditures stabilised at about 20 per cent of GNP in the mid-1960s. The current economic crisis has led to policies aimed at reducing governmental expenditures and deficits. To date, the effect of these policies has been to halt *increases* in expenditures, but not to reduce the total.

DEVELOPMENTS IN THE INTERGOVERNMENTAL SYSTEM

Nationalising IGR

It is difficult to identify separate phases in the development of the compound interwoven system of IGR in response to the foregoing pressures; it has been a gradual and ever-continuing process, generally characterised as one of centralisation.

As early as 1962 a government committee had warned against the centralisation of the system and erosion of local autonomy.[8] However, it would take until 1980 before the call for decentralisation became official policy, with the transfer of task and authority from 'higher' to 'lower' governments.[9] So far, the main result of this policy has been cynicism, and criticism of lack of substance. The few decentralisation plans which have been implemented (e.g. in the area of urban renewal) seem to serve mainly budgetary objectives.

The lack of results contrasts sharply with the general agreement over the problems of the intergovernmental system. Practitioners and academics accept that the system of Home Administration has become centralised to the point where costs in terms of efficiency, control, bureaucracy and effectiveness far exceed benefits. The original ideal of a decentralised unitary state is still shared, but its realisation has been overtaken by events and practices in the real world.[10]

The failure to implement the decentralisation policy has several causes. Observers point to the lack of coherence in the proposals: the impotence of the Ministry of the Interior in the power struggle with sectoral ministries which favoured national regulation over decentralised policy-making; the absence of political support at both ministerial and parliamentary levels; the resistance of nationally-organised pressure groups and professional organisations; the reluctance of national government agencies to give away power; and the inadequate 'decentralisation strategies' deployed by the Ministry of the Interior. Such an explanation of the 'failure' underscores implicitly the hierarchical image of the Dutch intergovernmental system. It looks only at resistance or failures *at the national level* and ignores the possibility that actors at the subnational level have a 'stake' in 'territorial politics'.

The evidence on the centralised nature of Dutch territorial politics and IGR is, however, unsatisfactory.[11] The data are aggregate and national in character and inapplicable if not irrelevant for understanding a disaggregated system and the role of the local level. By assuming that the structure is hierarchical, evidence about the interweaving of activities and levels of government is almost automatically interpreted as unilateral influence from 'centre' to 'periphery'. However, recent developments in the Dutch intergovernmental system have to be seen as a growth in complexity and interdependence. The classic image of the Dutch intergovernmental system is out of date: there is no longer a relatively simple structure of three layers of government, more or less independent from each other and engaged in their own affairs.

To avoid caricature, it should be noted that several observers in the 1920s and 1930s had concluded that the constitutional scheme of territorial decentralisation did not match the operation of the system of Home Administration. Although the scope of government activity remained restricted, in those areas where government actively participated in socio-economic life, a patchwork quilt of horizontal and vertical relations between different local, provincial and national ministries had started to emerge. For the public utilities, the image of three separate tiers of government was misleading.

The years after the 1914–18 War saw the development of 'collaborative decentralisation'.[12] The phrase refers to the development of multitudinous forms of co-operation among public and private organisations at different levels of government. For van Poelje,[13] a key characteristic of the 'modern state' is the combination of the nationalisation of legislation and rule-setting with administrative decentralisation for the execution of governmental tasks. This administrative decentralisation did not necessarily take the traditional form of territorial decentralisation. Rather than transferring tasks and authority to provinces and municipalities, they were vested in functional agencies, in which both governmental and non-governmental groups and organisations participated.

Similarly, at the local level van Poelje identified the emergence of functional decentralisation in the administration of municipal affairs. Many tasks were carried out by more or less independent organisational units. Often local governments subsidised non-governmental institutions to carry out specified tasks on their behalf. But if the multi-organisational, multi-actor characteristics of the Dutch system of IGR had been identified, they had not supplanted the traditional, centralisation thesis. The next challenge to this conventional image came with the post-war growth of the welfare state.

Welfare society

A central role in state affairs is played by private initiative and this has been described as the characteristic feature of the institutional structure of the Dutch welfare state.[14] From the 1930s the welfare state developed under the uninterrupted guidance of the confessional political parties which dominated government until the mid-1960s.

The conjunction of political dominance by confessional political parties and the development of the Dutch welfare state accounts for the importance of 'private initiative'. Confessional ideologies stress that government action in the public interest should be entrusted to private institutions, which are either already engaged in promoting those interests or are willing to do so.[15] The privatisation of governmental affairs is a well established principle in the Netherlands, and is a product of social and religious pluralism rather than economic (cost) considerations.

Given the influence of corporatist values and principles, a focus on the welfare *state* runs the risk of paying attention only to government institutions and overlooking the 'grey' area between citizenry and government. In this area, a plurality of social organisations and professional groups carry out delegated tasks for the state whilst meeting their own objectives and organisational imperatives for growth and power. Given the important role of social and religious groups and institutions in the overall structure, van Doorn prefers to talk about a welfare *society* instead of a welfare *state*.[16]

Until the mid-1960s the different social and religious groups and organisations were part of Lijphart's integrated system of 'pillarisation', but by the end of the 1960s the system of pillarisation had started to break down. The formerly integrated subgroups and organisations started to take on a life of their own. Some of them evaporated, but others were successful enough in terms of goal-displacement and adaptive growth to survive. How these structural changes in the political system affected developments in the operation of IGR remains an unanswerable question. With occasional exceptions, the consequences of de-pillarisation have been explored at the level of national government only.[17]

Despite van Poelje's early recognition of the importance of non- or quasi-governmental organisations in the Home Administration, it has not received much attention from scholars in the field of public administration. After 1945, attention was directed to the problem of reorganising governmental structures. Disaggregation and interdependence still did not occupy the centre stage.

Reorganising the territorial state

In retrospect, the process of trying to reorganise the territorial structure of the Dutch intergovernmental system began in the late 1940s, and demands for reform go back even further.[18] In 1947 a governmental committee — the Committee Koelma — recommended a fourth layer of government between municipalities and provinces. Districts were thought to be necessary in order to deal with the supra-local and regional problems which were alleged to cause many of the frictions within the intergovernmental system. The idea was rejected. In 1950 a law was enacted which facilitated intermunicipal co-operation and thereby eased supra- and inter-local problems.

During the 1960s, support for a separate regional level regained popularity. In 1969 a governmental note was issued proposing an extension of the Law on Joint Provisions (1950) so that municipalities could create general-purpose regional authorities (rather than limited purpose or functional authorities permitted by the existing law on joint provisions). Attempts in 1971 to make this proposal a law ran into strong parliamentary opposition. Fears of a fourth layer of governmental and the resulting complexity in the structure of subnational government played an important role in the debates. The proposal was withdrawn. However, the momentum for a 'rationalisation' and 'democratisation' of the system, for a fundamental reorganisation at the municipal level, had not dissipated. Proposals followed thick and fast. In 1974 the Ministry of the Interior issued a document proposing the division of the Netherlands into 44 socio-economic regions as the starting point for reorganising the structure of local governments. After a change in the governing coalition in 1975, attention shifted from municipal to provincial reorganisation. Twenty-six 'new-style' provinces were proposed to bridge the 'regional gap'. The tasks and responsibilities of these provinces were to be extended but, again, the plans gradually underwent several changes. Both the number of provinces and their proposed tasks were reduced. In 1977 the new provinces numbered 24; by 1978 the Minister of the Interior suggested 17, and in 1983 the idea of a provincial reorganisation was dropped altogether. The solution to the problem of inter- and supra-municipal frictions lies, once more, in the co-operation of municipalities. On 1 January 1985, a new Joint Provision Act came into force. The process of 'territorial reorganisation' had come full circle.

Municipal co-operation and amalgamation

Throughout the inconclusive search for structural reform the intergovernmental system continued to change. Under the Joint Provision Act (1950) inter-municipal co-operation continued to increase. Some limited forms of general purpose co-operation have developed. The bulk of co-operative arrangements are functional in nature. Depending on how one defines 'joint provision', the estimated number varies from 1,250 to 2,000.[19]

The number of municipalities has continued to fall, as national and provincial governments enlarge the scale of municipal government. Table 2 provides an overview.

TABLE 2
NUMBER OF DUTCH MUNICIPALITIES 1850–1984

Year:	1850	1900	1950	1960	1965	1975	1980	1984
Number:	1209	1121	1015	994	967	841	811	750

Source: Central Bureau of Statistics (CBS), 1984.

Table 3 provides a summary of the size (population) of municipalities.

TABLE 3
NUMBER OF MUNICIPALITIES BY POPULATION

NUMBER OF MUNICIPALITIES IN THE YEAR:

NUMBER OF INHABITANTS	1950	1965	1975	1980	1984
under 5,000	624	498	302	246	192
5,000–20,000	314	362	392	407	384
20,000–50,000	54	70	105	114	122
50,000–100,000	13	23	27	27	34
100,000 and more	11	14	16	17	17
TOTAL	1015	967	842	811	750

Source: Central Bureau of Statistics (CBS), 1984.

Central–local finance

Naturally, the development of the welfare society had an impact on central–local financial relationships. The scope of government activity expanded, and many of the tasks had to be implemented by the municipalities. Relationships between national and local governments intensified. Municipalities were not only assigned new tasks, but they were also confronted with various attempts to influence and control the way in which programmes and policies were conducted at the local level by central ministries. Besides intensifying provincial supervision, national ministries imposed various legal duties (e.g. planning) and addressed an increasing number of circulars to municipalities and other actors at the local level. These circulars contained guidelines, requirements, policy changes and clarifications about the way in which programmes and laws should be implemented. By the early 1980s roughly 1,000 circulars were sent to municipalities by the various ministries each year.

The growth of government also changed the composition of municipal revenue. Table 4 shows that in 1955, 50 per cent of total revenue was in the form of the general grant from the Municipality-fund. About 30 per cent of the revenues were allocated by specific or categorical grants. By 1960 this situation was reversed. At present, about 30 per cent of municipal financial resources comes from the general grant, leaving municipalities dependent on categorical grants for 60 per cent of their total income.

TABLE 4
SOURCES OF MUNICIPAL REVENUE 1955–85

Source of revenue	1955 %	1960 %	1965 %	1970 %	1975 %	1980 %	1985 %
1. Municipal taxes	14	11	8	7	4	6	9
2. General grants	56	41	41	39	34	31	27
3. Categorical grants	30	48	51	55	61	63	64
TOTAL	100	100	100	101*	99*	100	100

* Columns do not total 100 per cent because of rounding-up

Source: Teldersticklug, *Marke*, (The Hague: Staatsuitgeverij, 1985); J.L.M. van Wesemael, *Specifieke uitkeringen en de financiele zelfstandigheid van de gemeenten*, (The Hague: VNG 1982).

IGR IN THE NETHERLANDS: THE POLITICS OF HOME ADMINISTRATION

If one glances at the history of ideas about the Home Administration, one cannot be but struck by the fact that the typical political structure of the Netherlands – 'the politics of accommodation' – hardly ever plays a role in its analysis. Similarly, the structure of Dutch IGR is seldom taken into account by theories of the pluralistic nature of Dutch politics. The basic structure of the decentralised unitary state provides a ready complement to the need for consensus-building, mutual adjustment, accommodation and pacification. Yet these predominant features of the plural political system are divorced from Dutch IGR.

THE NETHERLANDS: A DECENTRALISED UNITARY STATE

Territorial politics: ideology and political parties

The role of party politics and political ideology in, and their effects on, Home Administration have not been studied thoroughly. Attention has been restricted to such topics as party political preferences for the reorganisation of the territorial structure; the appointment of Mayors and Queen's Commissioners; the analysis of regional economic policy; and the role of party considerations in the allocation of projects or resources to certain areas. Political geography is not seen as a very important dimension in Dutch national politics; a viewpoint reinforced by the lack of outspoken regionalist or separatist political movements.

The basic flaw with this interpretation is that it treats Parliament as some kind of a sovereign entity located outside or above the rest of the politico-administrative system. A Parliament does not need to be composed of representatives of regional, or local, levels of government to have territorial considerations, interests or viewpoints enter the decision-making process. Furthermore, the absence of regional-territorial political cleavages should not lead to the conclusion that subnational factors are either lacking or unimportant.

It is well-known that the national and local élites of a party often have opposing views on the same issues. The political preferences and preferred strategies of a Social-Democratic Alderman in a predominantly Christian-Democratic city located in one of the Southern provinces may differ from the preferences of his colleague from a big city in the Western part of the country, where Social-Democrats are the dominant party. Thus, a coalition of Social-Democrats and Conservatives has been impossible so far at the national level, but in several municipalities such ruling coalitions exist. Finally, if the law required that Burgomasters be elected by the local population the results would not be substantially different to the current system of national appointments.[20] Under current rules local considerations are taken into account in reaching a decision, revealing something about the interpenetration of national and territorial politics.

Dutch territorial politics are too varied to be reduced to one or two dimensions or categories. National policies and the programmes of political parties are not developed in a vacuum. There are many indirect and bottom-up ways in which local and territorial considerations enter the decision-making process. To what extent, and in what ways, these opportunities are actually used by municipalities remain to be seen. The topic can be studied adequately only if Parliament ceases to be the focus and becomes a locus within which the game of territorial politics is enacted. Parliamentary decision-making has to be seen not as the source, but as the end product of processes and activities within the broader politico-administrative system: Parliament is a compound actor in the midst of a multitude of organisations, groups and other forms of association. No unitary state's system of intergovernmental relations can be understood if it is treated as one big organisation, with the national government or Parliament at the appex of a line of command.

Territorial politics of IGR

One-sided, top-down steering and conflict resolution are less prevalent than one would expect for a 'hierarchical' unitary state. Many decisions *are* made at the national level of government, but — and the qualification is crucial — they are made by taking into consideration the possible effects on other levels of government. The results of top-down legislation bear the hallmarks of bottom-up re-enactment.

In important areas, such as physical planning, the legal framework typifies the decentralised unitary state.[21] Much to the regret of some planning theorists, central influence upon municipal land use plans is confined to provincial supervision, a relatively weak instrument for central steering. On several occasions, national plans had to be adjusted to municipal ideas and interest, and the resulting policy of 'the whole' became a compromise between the interests of the national and the local 'parts of the whole'.

Similarly, local governments have continuously resisted plans to strengthen the position of provinces in IGR. Inter-municipal co-operation was the long-standing preference of local governments over and against the preferences of the Ministry of the Interior and other national agencies, and it was the solution eventually adopted.

Finally, the top-down provisions — such as the power of national ministries to impose certain duties on municipalities, even against their will — are either not implemented, or the duties are carried out in quite different and unexpected ways. Such policy slippage has been explained as a product of 'the Dutch culture',[22] a phrase which means that one does not know how to explain it. The explanation lies, of course, in the structure and process of Dutch IGR.

The Dutch system of IGR is best conceived as a set of interorganisational networks. To analyse these networks attention must be paid to:

- the *organisational resources* that are deployed in intergovernmental relationships;
- the *institutional context* in which these resources are deployed; and
- the actual *political use* of resources, given the opportunities and constraints embedded in a particular institutional context.

National and Local Resources: Interdependence

In the relationship among national and local governments many different resources are of potential importance. Traditionally, most attention has been paid to the constitutional and legal aspects of the system. As already noted, the Dutch Constitution leaves the municipalities free to look after their 'own household'. However, there is no objective or generally accepted criterion for defining what constitutes the municipalities' 'own household'. Their autonomy can be defined only negatively, as that which is not being regulated by provincial or national governments. In addition, subnational governments are required to implement national laws and co-government has become a principal feature of central–local relations. Content analysis of co-government arrangements invariably shows a predominance of hierarchical provisions, formally enabling national authorities to intervene in local affairs.[23] Only rarely are national or provincial authorities formally dependent on municipalities.

The resulting hierarchical picture, however, needs substantial qualification. First, the top-down character of many legal arrangements is purely conventional. By definition and tradition laws have to spell out what does not belong to municipal autonomy. The Netherlands do not have a tradition of positively spelling out by an act of Parliament the bounds of municipal authority. Secondly, despite strict top-down legal provisions, the attitude of local governments, appear to be the decisive factor in explaining policy outcomes.[24] Thirdly, ministries use extensive legal powers only for a limited purpose. Their major aim is to keep the relative costs of local proposals as low as possible.[25] There is no need to explain this self-limitation in the use of formal powers by appeal to either central benevolence or respect for decentralist values. It is in the centre's interest. The centre knows it is dependent on the co-operation, information and local presence of other actors in the system. Hierarchical control and central intervention would foster reluctance bordering on recalcitrance.

Fourthly, the actual use of formal arrangements can be quite different from the centre's stated intentions. In many provinces, for example, supervisory powers are used to facilitate bottom-up co-ordination between adjacent municipalities, to exchange expertise and experience and for the province to play a consultancy role.[26] National field organisations, which are also meant to control and supervise municipalities, have similarly 'redefined' their role and willingly compensate for a lack of expertise within a municipal bureaucracy. Furthermore, informal agreements among municipalities, provinces and national field inspectors located in the region, render a legal obligation to get national approval, a time-consuming formality.[27]

Finally, the centre's legal powers of compulsion are often not used[28] or, on other occasions, are used to put a formal seal of approval upon a previously informally negotiated agreement among national and local governments.[29] There are also several examples of procedures designed by their legislators to be exceptions, but which have become daily practice among local and provincial governments.[30]

'The golden strings'

In recent years, central–local financial relationships have been the centre of attention. In 1980 the Council of Municipal Finances reported the existence of 532 specific grants for subnational, but mainly local, governments.[31] Government attempts to rationalise and simplify the system, by transferring money into block-grants and the Municipality-fund, had some success in its first year. In 1985 the Council of Municipal Finances observed with regret that the process had come to a halt,[32] and it blamed resistance at the national level as the main obstacle. In so doing, it overlooked, or at least underestimated, political support for specific grants at the subnational and local level.

Municipal officials are irritated by the bureaucratic and formalistic way in which national departments deal with local government. Such complaints are often ritualistic. The simple fact is that, in spite of local taxes and general grants falling as a percentage of total revenues, the decline is mainly the result of the *growth* in total municipal income and expenditures. Increases in public-sector expenditures have been a dominant feature of the development of the welfare society. The specific grant has been the instrument for passing part of this growth to local government.

Comparatively, Dutch municipalities spend a large proportion of total public expenditures.[33] Among EEC countries, between 1970 and 1981, Dutch municipalities were the second largest spenders after those of Denmark.[34]

Table 5 summarises developments in public sector expenditure by type of authority in 1950–84. In 1950, central government spent the largest proportion. From the 1960s onward, when total public expenditure started to increase sharply, the level of local expenditures rose above the national level. The greatest difference existed during the 1970s: the heyday of public sector expansion. Thereafter, under the influence of the economic crisis, the gap narrowed, in part because the national government had to cover social security deficits.

TABLE 5
PUBLIC SECTOR EXPENDITURE BY TYPE OF AUTHORITY 1950–84

Year	National Government (%GNP)	Subnational Governments (%GNP)	Social Insurance (%GNP)	Total public Expenditure (%GNP)
1950	17.5	10.0	4.0	31.5
1955	18.7	11.4	4.3	34.4
1960	14.0	13.6	8.3	35.9
1965	13.6	15.8	12.0	41.4
1970	14.0	17.7	16.5	48.2
1975	16.0	21.4	21.9	59.9
1980	19.0	20.2	24.1	63.3
1984	22.2	23.0	24.7	69.9

* Subnational governmental expenditure is mainly municipal expenditure
Source: Ministry of Finance, 1984

The predominance of co-government is generally interpreted as a centralising feature of Dutch IGR. However, just as important is the tradition of co-government which has embedded municipalities in IGR networks. Municipalities may depend on national grants, and financial relationships might be asymmetric, but there is no unilateral dependency relationship. Co-government creates the dependence of national government agencies on municipalities in specific policy areas: e.g. urban renewal, housing, physical planning, noise reduction and cultural affairs. Policy effectiveness depends on the co-operation and efforts of municipalities.

Furthermore, specific grants are a useful corrective to the general grant. Necessarily, the latter system has to be based on some uniform assessment of 'basic municipal needs'.[35] The 'objective' allocation of funds to municipalities cannot accommodate the variety of local needs.

Finally, the recent trend in central–local finances shows municipal taxes as an increasing proportion of municipal revenue, in part because of the decrease in resources from the Municipality-fund. The number of categorical grants has also been reduced, although the total volume of categorical grant aid allocated to municipalites has continued to grow (see Table 4).

One might interpret the reduction in general grant as increasing centralisation. Alternatively, it might be the case that it is easier for the Dutch government to reduce the amount of taxpayers' money allocated to the Municipality-fund and change (as the city of Rotterdam discovered) the formulas by which the money is distributed to municipalities. Such interventions are not centralising in intent, and their effects are general and diffused throughout the system. On the other hand, categorical grants are related to the specific interest of both governmental and non-governmental actors at national and local levels of government. Political support can be mobilised to resist (or reduce) potential cuts: opposition which is reported by the media. It is in the centre's interest to reduce the number of categorical grants. It could limit the number of potential, and perhaps unintended, political allies of local governments. It could also reduce the freedom of the municipalities to spend in absolute terms.

Political Support

In the analysis of Dutch IGR, organisational and political factors tend to be ignored. Conventionally, cleavages are between 'the Dutch municipality', the 'province' and – most often – the 'central government'.

In central–local disputes, the centre is seen as commanding the support of the electorate, pressure groups and political parties, and, judged by voter turnout, the Dutch voter seems more interested in national than in local issues. Pressure groups are organised on a functional basis, resist decentralisation and favour uniform national standards over local variation, while political parties are concerned predominantly with national issues and, for example, do not provide effective political support for the policy of governmental decentralisation. However, even if analysis remains confined to the national political environment, municipalities command important political and organisational resources. The Association of Dutch Municipalities (*Vereniging van Nederlandse Gemeenten* [VNG]) is one of the most powerful pressure groups in the Netherlands. All municipalities are members. It provides a variety of services and represents municipalities on all kinds of (functional) boards and committees. It is an important partner in central–local consultation and negotiation, invariably playing a central role in debates about problems and their solution.

Apart from the VNG many other associations of municipal professionals, public servants and local politicians (administrators) serve to exchange expertise and are, at the very least, important consultants for experts in and committees of the national bureaucracy.[36]

The distribution of political and organisational resources in the Dutch IGR network hardly justifies, therefore, the prevalent hierarchical interpretation. Many features of contemporary IGR serve the interest and objectives of actors at *all* levels of government. The evidence suggests that interdependence, not centralisation, is the main characteristic of the interwoven structure of IGR in the Dutch welfare society.

Institutional Context: Organisational Fragmentation

'The national government' and 'the municipality' do not exist: they are abstractions. Different levels of government should not be treated as if they were undifferentiated categories, for not only at the national levels but also at the local level there is extensive fragmentation. The big cities – Amsterdam, Rotterdam, The Hague and Utrecht – claim separate formal-legal status, and they have access to separate institutionalised channels of negotiation with 'the national government'.[37]

There are other kinds of functional differentiation. Municipalities from the various regions join functionally organised associations to exchange expertise and mobilise support at the national level. Coalitions are formed which cut across the central–local distinction and create the potential for conflict between different central–local coalitions. Within the locality, and despite the ideology of the 'unity of municipal administration', specialisation and segmentation are rife and have become major features of the politics and organisation of the larger cities. As already noted, institutional fragmentation generates costs in terms of time, negotiation, inconsistency and lack of clarity, and prompts the conclusion that the Netherlands are governed 'by thirteen disunited ministries'. This picture contradicts the image of a strongly centralised system of Home Administration: stated bluntly, if 'the national government' does not exist, then 'the power of the national government' becomes a myth.

Institutional fragmentation at the national level creates 'unexpected discretion' for municipalities.[38] The ideal of a rationally-integrated system of governance is pervasive, leading many commentators to underestimate the degree of fragmentation and interdependence. Yet municipalities play ministries off against one another to win resources, often for one and the same project. Multi-pocket budgeting[39] is also a well-known strategy among Dutch municipalities. More generally, it is worthwhile for municipalities to take the initiative in dealing with the fragmented national government system. They can exploit their superior local knowledge and exploit disagreements or simple lack of co-ordination among national ministries to get projects approved.[40]

Overlap and competition among national agencies enhances municipal autonomy. A fragmented institutional structure not only increases the influence of the policy sectors but also the organisational dependence of national ministries. For example, to avoid (additional) budgetary cutbacks, national agencies have to demonstrate the importance of, and need for, their programmes. To do so they depend on information, support, and 'fundable projects' generated at the local level.[41]

The failure to study this dimension of IGR makes it difficult to draw firm conclusions about the degree to which municipalities exploit the opportunities presented by institutional fragmentation. But the growing body of evidence on, for example, implementation suggests that the hierarchical-centralistic image of the Dutch intergovernmental system is grossly misleading.

Function and Territory

Functional differentiation within the intergovernmental system began with the development of the Dutch welfare society in the first half of this century. With the expansion of public sector activities, a further increase in specialisation and functional organisation was inevitable: functional organisation did not replace but supplemented territorial politics and administration. Territorial agents are constantly confronted with the demands of functional (professional) specialists, and the latter are, in turn, confronted by territorial actors (e.g. Burgomasters, Town Clerks, Queen's Commissioners, financial and planning departments at provincial and municipal levels of government). Professional demands have also to be adjusted to variations in local circumstances; to the varying size, structure and political composition of municipalities. On many occasions, therefore, sectoral specialists find 'their' policies modified; sectoral interests are traded-off against other local or sectoral interests (as in the case of physical planning). The current fragmentation of the politics of IGR is characterised by cross-cutting territorial and functional dividing lines.

THE POLITICS OF IGR: SEGMENTATION MEETS INTERDEPENDENCE

Resources and the constraints and opportunities which are embedded in a given institutional structure create the decision-making context. They provide an arena for political entrepreneurship. The actual outcomes of IGR depend on serendipity and other intangible factors, including the strategic capabilities of the actors, forms of leadership and the inclination and ability to play the game of bureaucratic politics. The role that Burgomasters play in this context has long been acknowledged,[42] although seldom studied. Recently, some former national political leaders have been appointed to the post of Queen's Commissioner at provincial level. Their overt lobbying to promote the interests of their regions is still regarded with some disdain. However, this change might stimulate long-neglected studies of this aspect of territorial politics. The combination of multi-dimensional fragmentation and the complex cross-cutting of functional and territorial cleavages explains the incremental nature of territorial politics in the Dutch welfare society. The system has developed and expanded gradually and steadily, without some 'Grand Design'. The first signs of stagnation in the welfare society were accompanied by warnings that it would be impossible to redesign and redirect the system in the same, incremental manner in which it had been built. The 'new design' is still awaited. The process of incremental, and sometimes drastic, adjustment is under way, but not without heated debates and political conflicts. The rate of change is slower than some wish and, as far as territorial politics is concerned, it is hard to identify either clear changes in style or key events.

An IGR-system which is characterised by functional and territorial fragmentation among interdependent actors needs shared, flexible and realistic rules of the game if political and administrative processes are to cope with change. The combination of political pluralism, institutional fragmentation and resources-dependence among actors in IGR-networks requires specific provisions for co-ordination, negotiation, conflict regulation, consensus and coalition-building to enable the parties to deal with their diverse and sometimes conflicting interests and relationships.

The key roles among IGR-participants in the Netherlands are characterised by complementarity, deliberation and 'power-free' consultation.[43] In short, organisational independence is associated with organisational insulation and protection against other, especially 'higher' authorities. The search for cut-backs in municipal expenditures has not altered this situation. Municipal expenditures have not been cut in order to reduce overall public expenditures. The Municipality Fund and ministerial budgets have been reduced, but the process of incremental budgeting has changed in content, not in nature. Of course, municipalities protest. But more often than not their anger is directed at the process rather than the cut-backs as such. They feel left out, neglected and unable to influence effectively solutions and strategies adopted by the national government. Ironically, an integrated, systematically developed and explicitly declared national government policy to deal with the reduction of municipal expenditures would probably be seen as a centralist intervention. But, at the same time, it would at least give municipalities something to fight against or demand influence upon.

The neglect of local interests at the national level and difficulties in reaching mutually-binding agreements among actors in the IGR-network are long-standing problems.[44] The tradition of insulated layers of government encourages voluntary relations; interactions frequently lack commitment. Interdependence among otherwise segmented units and subsystems, coupled with limited opportunities for mutually binding interactions, stimulate the search for predictability and efforts to stabilise and control (parts of) the environment. Many actors in IGR are, for example, in favour of planning. Close inspection reveals that this quest for planning refers, in fact, to the activities of *other* actors. It is a way of finding out the goals, motives and future activities of other agencies. When it concerns their own activities, planning is less attractive because it could make them more vulnerable to external control.[45] In these circumstances, planning, whatever its intrinsic merits, is more likely to increase the amount of paperwork than improve the predictability of the system.

There are many other occasions when the voluntarism of IGR increases the likelihood of formalism and deadlock. For example, the authority to approve plans is separated from the authority to fund plans. Approval does not imply funding. Furthermore, delays occur when a national government agency requires municipal plans in order to distribute and allocate money, but municipalities require information on future funding in order to develop plans.[46]

So long as specific grants are not based on the principle of 'payment for performance',[47] it is inevitable that national government agencies will seek ways of securing the interests for which they are held responsible by other actors in the IGR network (Parliament, pressure groups, labour unions, political parties and other national ministries). The available options include circulars, planning requirements, input criteria, supervision and inspection. Indeed, such forms of central steering, which have a stronger impact on the administrative workload of municipalities than on their performance, are not an effective means of central control.[48]

The recent trend is for 'covenants' among national and local agencies; 'policy-agreements' between the Cabinet and the VNG; and other more interactive, contractual modes of IGR to overcome some of these difficulties. It is too early to judge their effectiveness. In the meantime, and in retrospect, the main problem with Dutch territorial politics in the welfare society can most accurately be described as 'formalism' or 'bureaucratism' rather than 'centralism'.

CONCLUSION

The politics of Dutch IGR have been analysed conventionally, in terms of centralisation, control and local resistance. A hierarchical–centralist picture of the Home Administration emerged from a legalistic interpretation of IGR; a view reinforced by the nationalisation of the tax system and a corresponding dependence of local governments on national transfers. If the actual operation of the system is examined, however, it becomes clear that the picture is at best incomplete, omitting a range of formal, informal, structured, unstructured, organised and spontaneous politico-administrative processes. The outcomes of these processes are more open and less determined than the prevalent hierarchical perspective would suggest. The system is more accurately characterised as a set of IGR-networks; by interdependence rather than centralisation. The long-term and gradual development of an interwoven intergovernmental system is a product of increasing specialisation, professionalism and differentiation of state activities.

Van Doorn's prognosis[49] for the welfare society also contains lessons for IGR. He predicts that the state will become disconnected from society and that existing interrelationships will be untied. State interference will not be lessened; it will merely become different in nature. For IGR, this perspective suggests that the national and local levels of government will become disconnected, and the interwoven system will be unpicked. If the preferred means is decentralisation, central intervention will be frequent but take different forms. Disengagement will not involve detailed control or vague injunctions. Intervention may be more focused, harsh and hard-hearted. At the very time that the conventional picture of Dutch IGR gives way to an understanding of its complex of networks, the trend is towards simpler controls reminiscent of the simpler days of yesteryear. Whether or not the solution to the problems of complexity is to avoid them remains to be seen, but the costs of avoidance seem potentially high.

NOTES

1. Parts of this article are a modified version of Th. A. J. Toonen, 'Gemeenten en hogere overheden [Municipalities and Higher Authorities]' in W. Derksen and A. F. A. Korsten (eds.), *Lokaal bestuur in Nederland* [Local Government in the Netherlands], (Alphen a/d Rijn, 1985), pp. 350–64. The author would like to thank Hans Daalder, Rod Rhodes and Vincent Wright for their helpful comments on an earlier version.
2. The Water Control Boards can be seen as a functionally organised form of local government. They will not be taken into consideration here.
3. J. W. Fesler (1949), *Area and Administration* (Alabama: University of Alabama Press, 1964), p. 6.

4. J. H. Weggemans, *Gemeentelijk personeelsbeleid: een vergelijkend onderzoek naar autonomie en coordinatie* (Enschede: Technische Hogeschool Twente, 1981).
5. See, for example, Commissie Hoofdstructuur Rÿksdienst, *Eindadvies van de Commissie Hoofdstuctuur Rijksdienst* (The Hague: Staatsuitgeverij 1981).
6. P. A. H. H. Coumans, *Gemeentelijk inkomsten- en reinigings-beleid: verslag van een vergelijkend onderzoek* (Enschede: Technische Hogeschool Twente 1981).
7. E. F. Tufte, *Political Control of the Economy* (Princeton, NJ: Princeton University Press, 1978), pp. 98–9.
8. Commissie Territoriale Decentralisatie, *Rapport van de Commissie Territoriale Decentralisatie* (The Hague: Staatsuitgeverij, 1962).
9. Ministerie van Binnenlandse Zaken, *Decentralizatienota* (The Hague: Staatsuitgeverij, 1980).
10. M. Oosting, (ed.), *Aspecten van Decentralisatie* (The Hague: Staatsuitgeverij, 1984).
11. Th. A. J. Toonen, 'Administrative Plurality in a Unitary State: the Analysis of Public Organizational Pluralism', *Policy and Politics* (1983), Vol. 11, No. 3, pp. 247–71. Toonen, 'Implementation Research and Institutional Design: The Quest for Structure', in K. Hanf and Th. A. J. Toonen (eds.), *Policy Implementation in Federal and Unitary Systems: Questions of Analysis and Design* (Dordrecht/Boston: Nijhoff, 1985), pp. 335–54.
12. G. A. van Poelje (1942), *Algemene Inleiding tot de bestuurskunde* (Alphen aan den Rijn: Samson, 1953), p. 51.
13. van Poelje, *Osmose: een aantekening over het elkaar doordringen van de beginselen van openbaar bestuur en particulier bestuur* (Alphen aan den Rijn: Samson, 1931), and *Algemene ...*
14. J. A. A. van Doorn, 'De verzorgingsstaat in praktijk', in van Doorn and C. J. M. Schuyt (eds.), *De stagnerende verzorgingsstaat* (Meppel: Boom, 1978), p. 18.
15. S. W. Couwenberg, *Het particuliere stelsel: de behartiging van publieke belangen door particuliere lichamen* (Alphen aan den Rijn: Samson, 1953); D. Nichols, *Three Varieties of Pluralism* (London: Macmillan, 1974), p. 59.
16. Van Doorn, 'De verzorgingsstaat'.
17. However, see K. L. L. M. Dittrich, *Partijpolitieke verhoudingen in der Nederlandse gemeenten: een analyse van de gemeente-raadsverkiezingen 1962–1974* (Leiden: Rijksuniversiteit, 1978).
18. J. In't Veld, *Nieuwe vormen van decentralisatie* (Alphen aan den Rijn: Samson, 1929), p. 218.
19. Toonen, 'Bestuurlijke reorganisatie door gemeentelijke samenwerking', in W. Derksen and A. F. A. Korsten (eds.), *Lokaal Bestuur in Nederland* (Alphen aan den Rijn/Brussels: Samson, 1985), p. 321.
20. W. A. Derksen, *Monumentenzorg en effecten van een centraal beleid, een analyse van de bescherming van stads- en dorpsgezichten* (Deventer: Kluwer, 1983); and Derksen, *De gekozen burgemeester benoemd* (Deventer: Kluwer, 1983).
21. P. Glasbergen and J. B. D. Simonis, *Ruimtelijk beleid in de verzorgingsstaat; onderzoek naar en beschouwing over de (on)mogelijkheid van een national ruimtelijk beleid in Nederland* (Amsterdam: Kobra, 1979).
22. Tj de Koningh *et al.*, *Ordening van de besluitvorming over de ruimte* (Deventer: Kluwer, 1985), p. 186.
23. F. Fleurke, *Invloedsverhoudingen in het openbaar bestuur* (The Hague: Staatsuitgeverij, 1983), p. 70; and D. W. P. Ruiter, 'Ontworpen Verticaal Machtsevenwicht', in Ruiter (ed.), *Verticaal Machtsevenwicht in het binnenlands bestuur* (The Hague: Staatsuitgeverij, 1983), p. 90.
24. Derksen, *Monumentenzorg*.
25. H. Bleker and W. H. van den Bremen, *Macht in het binnenlands bestuo* (Deventer: Kluwer, 1980), p. 90.
26. Weggenmans, *Gemeentelijk*.
27. H. J. A. M. van Geest, *Bestuurseffectrapportage II: een instrument ontwikkeld* (The Hague: Staatsuitgeverij, 1986).
28. De Koningh, *Ordening van de besluitvorming over de ruimte*, p. 198.
29. Derksen, *Monumentenzorg*, p. 165.
30. B. C. Brocking and van Geest, *Anticiperen en Ruimtelijk Beleid* (Zwolle: Tjeenk Willink, 1982).

31. Raad voor de Gemeentefinancien, *Overzicht specifieke uitkeringen 1980 die voor gemeenten en provincies van belang zijn* (The Hague: Uitgeverij van de VNG, 1980); Raad voor de Gemeentefinancien, *Heiligt het doel alle (specifieke) middelen* (The Hague: Uitgeverij van de VNG, 1981).
32. Raad voor de Gemeentefinancien, *Advies Gemeentebegrotingenf 985–1986* (The Hague: Staatsuitgeverij, 1985), p. 2.
33. R. Paddison, *The Fragmented State: the Political Geography of Power* (Oxford: St. Martin's Press, 1983), p. 34.
34. G. van Daele, *Evolution of Local Government Receipts and Expenditures in the EEC* (Maastricht: European Institute of Public Administration, 1984).
35. De Koningh, p. 230.
36. P. W. Tops and Korsten, 'Categoriale organisaties van gemeente-ambtenaren' in Korsten and Derksen (eds.), *Uitvoering van overheidsbeleid: gemeenten en ambtelijk gedrag belicht* (Leiden/Antwerp: Stenfert Kroese, 1986), pp. 123–35.
37. Union of Dutch Municipalities (VNG), *Agenda Consultation, Central Government, Four Large Municipalities* (The Hague: Uitgeverij van de VNG, 1979).
38. Derksen, *Munomentenzorg*, p. 164.
39. D. O. Porter and D. C. Warner, 'How effective are grantor controls?', in K. E. Boulding (ed.), *Transfers in an Urbanized Economy* (Belmont: Wadsworth Publishing Co., 1973), p. 287.
40. Bleker and van den Bremen, *Macht*.
41. J. L. Pressman, *Federal Programs and City Politics* (Berkeley: University of California Press, 1978), pp. 152–3.
42. H. A. Brasz, *Veranderingen in het Nederlandse communalisme: gemeentebesturen als elementen in het Nederlandse stelsel van sociale beheersing* (Arnhem: Vuga-boekerij, 1960), p. 168.
43. H. Bleker, *Na(ar) goed overleg* (Deventer: Kluwer, 1984), p. 47.
44. Vakgroep Bestuursrecht en Besturrskunde, *Verkenningen in verticale verhoudingen* (Groningen: Rijksuniversiteit, 1977), p. 97; Toonen, *De pluriformiteitsgedachte in het openbaar bestuur* (1982), p. 110; Van Geest.
45. Vakgroep, pp. 108–10.
46. Veldkamp Marktonderzoek, *Evaluatie van de uitwerking van het convenant* (Amsterdam: 1980), p. 23.
47. Pressman, *Federal Programs*, pp. 152–3.
48. VNG/SGBO, *Bestuur per circulaire* (The Hague: VNG-studies, No. 6, 1985), p. 49.
49. J. A. A. van Doorn, 'Anatomie van de interventiestaat' in J. W. De Beus, and van Doorn (eds.), *De Interventiestaat* (Meppel/Amsterdam: Boom, 1984), pp. 9–24.

Ireland: The Interplay of Territory and Function

T.J. Barrington

In March 1943, in a famous Saint Patrick's day broadcast, Eamon de Valera as *Taoiseach* (Prime Minister) chose as his subject 'The Ireland That We Dreamed Of':

> ... let us turn aside for a moment to that ideal Ireland that we would have. That Ireland which we dreamed of would be the home of a people who valued material wealth only as the basis of right living, of a people who were satisfied with frugal comfort and devoted their leisure to the things of the spirit − a land whose countryside would be bright with cosy homesteads, whose fields and villages would be joyous with the sounds of industry, with the romping of sturdy children, the contests of athletic youths and the laughter of comely maidens, whose firesides would be the forum for the wisdom of serene old age. It would, in a word, be the home of a people living the life that God desires that man should live.[1]

In other words, he evoked the kind of Ireland, rural and traditional, that de Valera himself grew up in at the end of the nineteenth century. Mr de Valera, a master politician, did little that was unpremeditated, and one wonders at such nostalgia when the Irish people were about to stumble, blinking, into the bright light of the post-war world − and focus a firm gaze on the country's material goods. In the process they were to accept significant change in their own values and attitudes.[2] Modern Ireland, although the product of the nineteenth century, is in the process of rapid, if selective, change, development, decline, depending on one's viewpoint.

NINETEENTH CENTURY FOUNDATIONS

At the end of the seventeenth century a new colonial ascendancy replaced all but a few remnants of the old ruling class, and the 'new men' were cut off from the great mass of the people by wealth, religion, political loyalty and power. In the nineteenth century four main forces brought about a new social equilibrium that dominated the independent Irish state almost to the present. These forces were the collapse of the seventeenth-century Protestant, landed ascendancy; the rise of popular, largely rural-based parliamentary democracy; the creation of a middle class; and the evolution of interventionist government.

The Protestant Ascendancy was undermined − except, in part, in the north-east − by the Union with Great Britain in 1801, the development of the franchise, the reform of local government, and the growth of central government. The agrarian disasters of the century, especially the Great Famine of 1845-47 and the lesser but severe one of 1879-80, financially destroyed many of the great land-owning families.[3] Politics in the last two decades of the nineteenth century finally destroyed what was left after economic decline.

Early in the century Daniel O'Connell mobilised the rural poor (Sean O'Faolain aptly named his biography of O'Connell, *King of the Beggars*),[4] to achieve Catholic emancipation in 1829. A later campaign, with its 'monster rallies' of up to a million people, failed to achieve repeal of the Union, yet O'Connell made a wide European impact. Thus were fused together the roots of Irish popular, rural democracy and of European Christian democracy.

Charles Stewart Parnell, a landlord and scion of an Ascendancy family, mobilised the disciplined masses in a heady, populist combination of nationalism and agrarianism. The first was again to fail; but the second had almost complete success, and a nation of peasant proprietors was born, so reinforcing a popular belief in the moral superiority of the rural way of life.

There was a small Protestant middle class — in trade, the professions and the public service; but only in trade were Catholics numerous. Their rise paralleled the rapid growth of secondary level education as the century wore on, almost wholly organised by the Roman Catholic Church. From mid-century the state supplemented this by establishing four university colleges. Steadily the professions were colonised, as the trader's (or strong farmer's) son became a doctor, priest or lawyer and, later, public servant or politician. Parnell, who died in 1891, was, outside the north-east, the last Protestant leader of any major Irish party. As the rural Ascendancy faded its place was taken by this breed of 'new men' whose basically conservative and Victorian values were to play a big part in forging the moral and political climate of independent Ireland. Yeats, in his '1916', captures the unexpectedness of their further rise to heroic patriotism. Rural values, based on numbers and assumed moral superiority, were long dominant; and with bourgeois conservativism there was little conflict. At least until recently it was part of the task of skilful politicians to avoid or resolve confrontation. Hence there existed a long-standing and conservative coalition, complacent in facing change and quite remarkably unintellectual. As the population became, as now, two thirds urbanised, and found their roots receding into time, as Victorian values and assumed religious certainties have begun to erode, this conservative alliance has begun to break down. There is a strong trade union movement — about two thirds of Irish employees are unionised — especially strong and militant among white-collar workers; but the working-class contribution — and still more that of socialism — has been notably small.

It was not until the 1830s that genuine attempts to govern Ireland were made. Thereafter, intervention increased apace. To take a few illustrations: from 1831 a national system of (primary) education emerged which was national at least partly because of the high level of religious tension at that time; from 1836 a national police force was created and was national for security reasons; from 1838 there was the application of the English Poor Law to Ireland, with its mixture of democratic local bodies and intense central control; and from 1840 the reform of the corrupt municipal boroughs was undertaken. In addition, a large number of essentially English measures were applied to Irish conditions, culminating in the abolition of the last bastion of the Ascendancy, the grand juries, and their replacement by popularly elected county and rural district councils. And, of course, Land Purchase Acts, that finally 'gave the land to the people', were introduced.

Government was forced into intervention by the logic of the Union and by the pressing necessities of Irish conditions, but was unable to decide between a centralised and a decentralised system. An extensive system of Irish local government was established to cope with health, urbanisation, roads and the like, but it had no responsibilities in the fields of education and policing. Central supervision was very tight over health and welfare and very loose over the other functions. This ambivalence did not trouble the new bourgeois rulers of twentieth-century Ireland, to a man democratic centralisers. The centripetal tendencies were, in the twentieth century, reinforced by the fact of *homogeneity* and by a lively consciousness of *smallness* and *poverty*.

The Partition of Ireland, whereby the six north-eastern counties remained in the United Kingdom, but with a form of Home Rule, and the remaining 26 counties became independent, left the population of the new state remarkably homogeneous, socially and religiously: it was over 93 per cent Roman Catholic. The one significant cleavage was caused by the post-independence Civil War of 1922–23; but even here the worst wounds soon healed. It was natural, then, to think of the state as a single homogeneous entity without territorial or corporate interests to be placated.

The island of Ireland is not all that small – some 500 kilometres long and about half that in width and 84,000 sq.km in area. The boundaries of one of its bigger counties are as far apart as a journey from Brussels and all but one of the capitals of the six founding countries of the European Community. But the population *is* small – some five million in the whole island, of which just three and a half million are in the Republic.

The Republic with a gross domestic product at market prices in 1984 per head of 6,740 US dollars is – just – in the top sextile of the countries of the world; but, until Greece joined the European Community, Ireland was its poorest member. So, in the 1920s there developed a lively sense of the need to keep things simple, small, economical, centralised; function rather than territory[5] was to be the basis for such limited government as would be needed.

Events were to belie many of these assumptions. Occasionally the political system might show some awareness of the need for new thinking, but such temptation was readily resisted. The nineteenth century, for all its disasters, was a century of re-birth and of creativity that made a lasting mark on the Irish consciousness and system of government.

PERIPHERAL POLITICS

From nearly every relevant *external* standpoint, twentieth-century Ireland is peripheral – to the United Kingdom, the European Community, the United States, the Roman Catholic Church. Internally, there are the overlapping pereipheries of Northern Ireland and some socio-economic and cultural peripheries, chiefly in the west of Ireland, and the socially marginalised groups.

IRELAND: THE INTERPLAY OF TERRITORY AND FUNCTION   133

For the 121 years during which Ireland as a whole was part of the United Kingdom, it was politically, economically, religiously and, to some extent, culturally, distinctive. After Independence in 1922 there was some political friction with the United Kingdom, as political ties were steadily loosened; but economic, social, cultural, and legal ties remained close − a common labour market, a common currency, each a major trading partner of the other, much cultural interchange, and a good deal of coming and going at the political, administrative and personal levels. The big conflict of interest was in relation to the effects of Britain's cheap food policy on a major food exporter.

Membership of the European Community − which because of the close trading links with the UK, became possible only in 1973 with the latter's accession − was widely welcomed. Among other things it has broadened trading relations: in 1970 62 per cent of Irish exports went to, and 50 per cent of Irish imports came from, the United Kingdom,[6] whereas in 1984 the percentages were 34 per cent and 43 per cent respectively.[7] The currency link with the UK has been broken, and a common market in currency, goods and labour is now becoming a Community reality. Ireland is the only member of the Community which is neutral and not a member of NATO. Notwithstanding the financial advantages to Ireland of the common agricultural, regional and social policies the gap between the richest part of the Community and Ireland in terms of gross national product per head has grown.[8] There is support, tempered by a wary concern, for the Community, but the historic drive for a European union has yet to capture the Irish imagination.

Because of heavy emigration, the homeward flow of emigrants' remittances, and the American support for Irish independence many parts of Ireland in the nineteenth and early twentieth centuries were psychologically more on the periphery of the United States than of the United Kingdom. The Irish Constitution of 1937, the Supreme Court and constitutional law have been heavily influenced by American constitutional jurisprudence. Emigration to North America is now much reduced; but some 30 million Americans claim Irish descent, and Irish and American politics occasionally mingle. A notable example of the latter was in 1983 when a Catholic movement to amend the American Constitution to ban abortion led to a similar movement to amend the Irish Constitution. And, of course, the troubles in Northern Ireland create some stir in the United States. Numerous Americans have been discovering, and in some instances cultivating, Irish roots, something that may be observed of Canada, and perhaps in the still limited two-way traffic with Australia. If location on the UK periphery bred paternalism then the American link supplied a touch of motherhood − or grandmotherhood?

The population of the whole of Ireland is 75 per cent Catholic (95 per cent now in the Republic) with a very high level of practice. There has been notable missionary activity in English-speaking parts of the world and, now, in many Spanish-speaking countries; but the Irish contribution to the intellectual life and central government of the Church has been limited.[9] For Ireland, immersed in the nineteenth-century ultramontane movement, much of what happened at Vatican II was a surprise; yet the changes were accepted with astonishing aplomb, showing a society prepared to move with practice, if not with ideas. (There is a joke that, when Pope John Paul I died unexpectedly, a last-ditch attempt to ensure a European successor required a move to Poland because of the inadequacy of the Irish telephone system!).

The extensive involvement in missionary activity has created much empathy with the problems of the Third World, as notable personal response to famine appeals has so vividly illustrated. Internally, the smallness of the religious minorities (only 5 per cent of the population of the Republic) means that they have little political weight should religious interests clash. The real problems have arisen from the efforts of extreme Catholic factions to bring about some sort of confessional state and the varying skills and levels of courage with which politicians and government have handled issues such as censorship, the Spanish Civil War, constitutional recognition of the churches, the welfare state, education, contraception, abortion and, most recently, divorce. There has been much shuffling, hanging back, backtracking and the occasional disaster[10] but, overall, there is a movement towards a modest pluralism. Here, the courts, in interpreting the civil rights embodied in the Constitution, have played a notable part. However, where the Constitution itself makes a specific prohibition, as it does with divorce, there is no alternative, if it is to be removed, to political decision, legislation and referendum — a process that in June 1986 failed when the professional politicians were hopelessly outclassed by amateur defenders of the status quo. The climate of the Republic may not be openminded but, in an absentminded way, it is tolerant and relaxed.

Because of proportional representation Irish governments normally have slender majorities; and increasing urbanisation makes political loyalties more volatile: hence the political importance of the more traditional, and more politically stable, areas of the West. These are also poorer areas. Although there has been convergence the gap is still considerable.[11] So, where patronage is feasible — extensive patronage is *not* a feature of Irish politics — the West gets special attention. For a generation or more, this tactic has tended to yield political dividends. Within the West itself there is another periphery, the *Gaeltacht*, in which the Irish language is spoken as a steadily shrinking vernacular, now a scatter of communities mainly confined to the western and south-western seaboards. A long-standing aim of Irish government has been to 'save the Gaeltacht', but adverse economic, social and cultural forces are strong while political commitment is weak. Industry, fishing and economic activity generally have been promoted, but by employing the general Irish principle of exogenous development: that is, what can be done *for* people and communities rather than *by* them. Dissatisfaction with this led to the rise, in the 1960s, of a Gaeltacht Civil Rights movement to establish a special Gaeltacht local government unit with wide developmental discretion.[12] After some delay, a new Gaeltacht authority was set up on the foundation of an existing ad hoc state-sponsored body, with elected members enjoying a bare majority of the board. The political parties moved in; the board became highly politicised, and the civil rights movement lost its bite. Irish politics skilfully copes with Irish centrifugalism.

Rather more than half of public expenditure is social expenditure; but successive governments have been slow to recognise the welfare claims of those outside the traditional categories — the travellers, the homeless, the lonely, the aged, the handicapped, and others such as those with broken marriages.

The more middle-class groups, where they are skilfully led, carry some political weight; but more economically deprived groups have had a tougher time. The notion that welfare requires government to seek out and care for those peripheral groups who have fallen through the existing official nets, and who are so deprived as not to be vocal, makes slow headway.

In Northern Ireland there are two major 'traditions', with some 60 per cent of the population of 1.5 million being 'British' and 'Protestant', and the other 40 per cent 'Irish' and 'Catholic'. However, identifiers with the 'British' tradition also regard themselves as 'Irish'. These differences, far from acting as stimuli for local unity, cause deep polarisation. Basically, in Belfast and its hinterland, the seventeenth-century settlement survives – the notion of the Ascendancy of the Protestant settlers from Scotland and Northern England, supplemented by later immigrants from these areas, now resting on solid working-class support and operating on the principle that, in scarce matters, 'winner takes all'. This is partly moderated by the equalising effect of the welfare state. The present troubles originated in the rise of a new, well-educated group among the minority, their agitation for equal civil rights, and the violent reactions, institutional and personal, to that movement. The United Kingdom government has guaranteed the majority that they will not be coerced into a United Ireland. It maintains a substantial security presence and, in the interest of equal benefits for all the citizens of the United Kingdom, provides substantial annual subsidies, estimated at £1,500 sterling per head of the population. It has also been trying to bring about *some* local political movement.

The Irish government is also unwilling that the island be politically re-united against the will of the Northern majority; but urges that the serious grievances of the nationalist community be tackled. It argues that the strife of recent years has exacerbated local differences; that it led to serious security problems within the Republic, now happily moderated; that (to put it crudely) the direct aggregate extra cost per head for security is four times as great for the Republic as for the United Kingdom; and the indirect costs – particularly in the damage to industrial investment, to tourism and to the general image of the country – all constitute heavy burdens that give it a direct interest in seeking to ensure that there is steady movement towards peaceful solutions.[13] While recognising that a number of reforms have been introduced, the Irish government has been pressing for reform in the judiciary, in the area of civil rights (especially access to employment) and in relation to certain sections of the security forces.

Although the two governments have a common interest in the Northern Ireland 'problem' the United Kingdom government has been reluctant to recognise it. However, the Hillsborough Agreement of November 1985 established a formal consultative structure between the two governments which is supported by a joint, continuing secretariat. The Agreement aims to examine means for improving joint security, for removing alienation of the nationalist community and for moving towards some generally acceptable form of devolution. Some improvement has been made on the first issue but, by late 1986, not on the latter two. The Protestant population bitterly resent the Agreement: the Catholic population, having seen no benefit to them from it, view it in sullen silence. It is too soon to judge whether there are, or are not, the ingredients for political movement here. The only hope for a civilised solution is, somehow, to get the political system to move. The essential conditions seem to be: continued sophisticated co-operation between the two governments; acceptance on all sides of how other societies, by recognising their divisions, can achieve a viable society; and, among the majority, of the responsibilities, and limitations, of numerical strength.

DEMOCRATIC CENTRALISM

While government under the Union could not make up its mind between a centralised or a decentralised system, the new government of independent Ireland had no such doubts. There was a consistent commitment to democratic centralism achieved by three main, interacting policies – for intense multi-function centralisation, for single function delegation, and for local subordination. Inner contradictions between these policies took time to emerge; in the meantime there was a practical job to be done.

Multi-Function Centralisation

The state established in 1922 was consciously and deliberately a highly centralised one. The revolutionary government that took over was determined to show itself fully in control of the system of government and to establish its willingness and capacity to rule, faced as it was with: a bitter Civil War with former comrades; well developed public institutions fully staffed by public servants transferred from British rule whose loyalty to the new order they naturally, if wrongly, distrusted; a major post-war slump; the general consciousness of the smallness and poverty of the whole; a 'business-like' determination to keep things simple, solvent and, so far as possible, depoliticised. The simplest expedient seemed to be to concentrate the business of government as far as possible in a small number of multi-function organisations. So, of the 47 official boards and offices at Independence the functions of all but two were concentrated into eleven ministries and, with minor exceptions, all power within the ministry was concentrated in the minister personally. This concentration was achieved by the device of making him, not the ministry, the legal corporation: he would have to take all important decisions, and many that were not. By centralising power and action in ministers, individually and collectively, they could be held clearly accountable to the Dail, the directly elected House of Parliament. 'Business-like' at that time meant almost exclusively small business, indeed one-man business, and that the daily conduct of public affairs should be kept 'above politics'.

The business of government steadily grew, and with it the size and number of civil service departments and agencies. The largest of these had regional and local offices but the system precluded the granting of discretion to senior officials in those offices: the seat of decision remained in Dublin. So, while these offices were located in regional and local areas they were not part of them and had no mandate to work with other public bodies operating in the same areas. There was no one, prefect or councillor, to knock bureaucratic heads together in the interest of co-operation and joint action with other civil service bodies, with state-sponsored bodies, and with local government bodies.

Hence growth, in the context of such a degree of centralisation, led to proliferation and fragmentation – and extensive frustration. From the late 1960s timid attempts were made, through regional consultative bodies, to remedy the worst ills of this system, with some modest success. But the basic centralised structure remains intact.

Single-Function Delegation

The one-man business model, in the shape of the multi-function centralised ministry, was barely in place when the first signs of breakdown began, with the emergence in the mid-1920s of functional delegation to the single purpose state-sponsored, or parastatal, body. They are now a most distinctive feature of Irish public administration. They total about 100, ranging from classic public enterprises, through promotional organisations, to various kinds of regulative and cultural bodies. Here the model, on which several variations have occurred, was to become the limited company under a ministerially appointed board of directors. These boards were free, and expected, to exercise a wide range of initiative and discretion. The discretion granted to these *appointed* representatives contrasts strongly with the extent to which the locally *elected* representatives were firmly tied down, contaminated, as they were, by 'politics'. The contradiction between the tight central control applied to the civil service and to local government, on the one hand, and, on the other, the 'business' freedom given to the state-sponsored bodies – although an increasing number, and now a majority, of them are providing services of a comparable nature – was never resolved and for long never authoritatively posed.[14] As the number of state-sponsored bodies steadily increased the accumulating issues of roles, of communications and central co-ordination were neglected.[15]

The functional agencies could count on a wide measure of quiet public approval, now diluted by recent spectacular lapses by some commercial state-sponsored bodies and by evidence of very bad decision-making within them. If this could happen to bodies subject to the rough but revealing discipline of balance sheet and profit and loss account, what about those other state-sponsored bodies subject to no such discipline, financial or political? Methods of central control, supervision and accountability are being tightened, so raising the question whether the system is entering a new cycle of multi-functional centralisation.

Local Subordination

The new, reformed and democratised local government institutions established during the nineteenth century played a significant part in the subsequent struggle for independence; nevertheless, their roots remained shallow. The inner force of Irish government, strongly influenced by British practice but lacking British countervailing centrifugalism, was almost wholly centripetal. In consequence, peripheral politics impinge on national institutions: there are no authoritative local institutions. There are three underlying reasons for local subordination; reasons that constitute a consensus. First, there is a belief that government should, wherever possible, be kept 'businesslike'. Second, centralisation and central control are seen as essential means to efficient operation. Third, it is deemed crucial that 'politics' be confined to the national stage and banished from the functional and territorial ones.

On Independence, many small elected bodies were swept away and their functions concentrated in 27 county councils and four county boroughs. Small town authorities survived but with increasingly irrelevant functions. Basically, and to achieve local multi-function centralisation, Irish local government was concentrated into 31 'county' authorities, the number of councillors was reduced, functions were limited, the autonomy and scope for initiative and innovation were denied.

To raise the professionalism of the local service recruitment to all significant posts was given in 1926 to a civil service-type Local Appointments Commission. And to raise the quality of management, first city managers and then county managers were introduced, again on the business model. Most stringent administrative control was exercised by the central ministries. The aim was to create a tight, effective system of local administration; that once achieved, central government lost interest. For the past 40 years local authorities have lost functions: their 'market share' has declined. The current ratio of local government expenditure to GNP is about 10 per cent.

Apart from housing, the welfare state made little lasting impact on local government. Education, for example, was overwhelmingly a central service, and from 1953, income support became wholly the task of a central ministry. From 1970, health − about half of the activity of local government − was transferred to regional bodies. The basis of local taxation was neglected. Notwithstanding warnings that the method of valuing buildings and land for local rates, a central service, had lost legality and equity, nothing was done. This neglect was rendered all the more reprehensible because the rate *collection* system, administered locally, was highly effective and efficient. Political and legal failure came at a time when the exchequer was incurring heavy deficits, and some 6 per cent of public revenue was simply 'lost'. Central government promised, but did not always deliver, compensation for local authorities for the loss of revenue − in effect, it coped with the tax revolt by adding to its own deficit! In return and inevitably, it acquired statutory control over local expenditure. Subsequent attempts to raise taxes from houses, to 'tax the farmers' and to charge for local authority services, reached depths, fortunately rare, of political and administrative incompetence.

The equanimity of public opinion on these matters was undisturbed. Thus, in 1969, when a Minister for Local Government installed a temporary commissioner in place of the Dublin city council, the oldest and largest Irish local authority, the public silence was deafening. Local councillors have considerable democratic legitimacy: voter turnout in the seven local elections between 1950 and 1985 averaged 62 per cent. Yet councillors have taken such treatment with little protest. As representatives of the periphery, councillors, neither collectively nor in their parties, carry much weight. Yet, relieving them of virtually all responsibility has opened the primrose path to democratic irresponsibility.

The depoliticisation of *local* government and the emasculation of *local* politics is almost total. Local politics are little more than the local struggles of the national parties; most councillors belong to a party and accept tight party discipline. Apart from the single issue of county management, there was hardly a murmur of complaint from councillors as their role was eroded. The absence of local politics and discretion and the extent to which all power was vested in Dublin inhibited the emergence of 'notables'. One's local standing depended on the local effects of achieving power at the central level. History had seen to it that there were no hereditary 'notables' and virtually no county élites. There is a good deal of county loyalty, even patriotism, but it takes almost exclusively a cultural, not a political, form. Centralisation of administration was paralleled by an intense centralisation of politics.

Cumul des mandats is a striking feature of Irish government. Some 60 per cent of existing parliamentarians are, or have been, members of local authorities, serving on sub-county, county, regional, and joint authorities. Indeed some are also members of the European Parliament and the Council of Europe. The Irish system of proportional representation depends on multi-member (3–5) geographical constituencies. Competition between and *within* the parties is intense so that there are virtually no 'safe' seats. One must, therefore, keep oneself continually before the public by occupying as many public positions as possible to eclipse one's rivals. And then there is the enormous – and apparently obligatory – chore of 'brokerage'. Locally, indeed nationally, the public representative may not have much power, but he has plenty to do!

The ramshackle administrative system is held together by an astonishing level of political clientelism, or 'brokerage', with its roots in an Irish cultural phenomenon – the widespread, deep (and unfounded) distrust of the honesty of government. Brokerage is promoted by the competition between parliamentarians to help their constituents whose (often minor) difficulties necessitate 'having a word with the Minister' or tabling a parliamentary question.

One effect of administrative proliferation has been to bewilder citizens and the political system itself. The great rise in the number of parliamentary questions has resulted in the breakdown of the system of oral reply. The rise in the volume of 'representations' to ministers (one unpublished 'guesstimate' suggests they may reach as much as seven figures annually) has led to much diversion of administrative activity. It is pathetic to see public representatives tumbling over themselves to claim credit from a constituent awarded a grant to which he is statutorily entitled; or whose daughter is to be appointed to a public service job won in a scrupulously conducted open competition.

Whatever the failings of the administration – principally a lack of skilled day-to-day management and undue tolerance for sluggish systems – it tries to serve the people honestly. However, size, proliferation and brokerage increase public distrust. Little attempt is made to tackle this distrust either by bringing the administration closer to the people by simplifying its bewildering complexity – the social welfare system has 36 different rates of child benefit! – or by providing adequate systems for advice and redress utilising advice centres, 'one-stop shops' and the rest. Typically, when something *had* to be done in the matter of redress, the smallest, most cosmetic step possible was taken: in 1984 an Ombudsman with very limited functions was appointed. But no one in authority is concerned for a *system* of orderly hearing and redress of grievances and for ensuring that such tribunals as exist cause no anxiety.

THE SYSTEM OF GOVERNMENT

The argument to date is that the undue pursuit and practice of centralisation has led to proliferation and fragmentation at the centre, stagnation at the periphery, and, linking the two, the distortions of 'brokerage'. Have there been adverse effects on the practice of government?

Figure 1 gives a simple outline of the Irish system of government as it has developed. There are over 400 governmental bodies, employing some 27 per cent of the total workforce. Government has grown absolutely and relatively:

FIGURE 1   THE IRISH SYSTEM OF GOVERNMENT

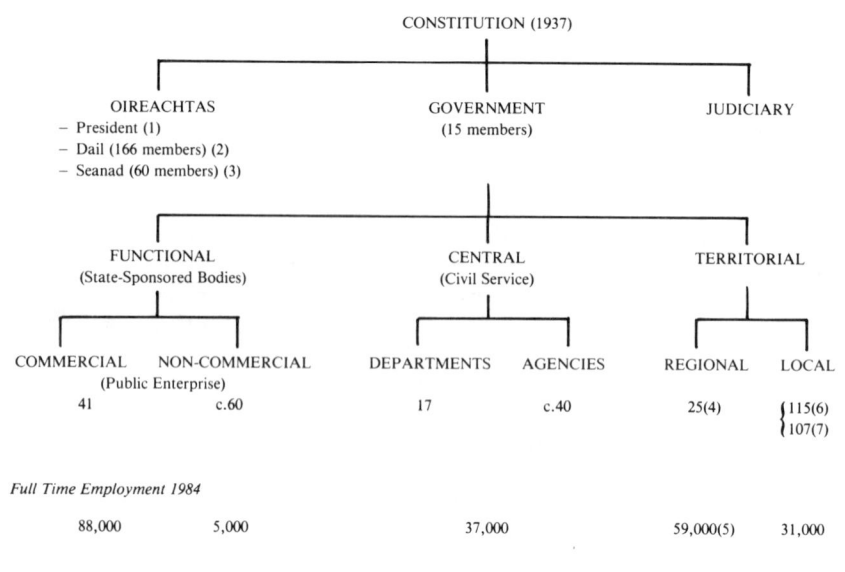

CONSTITUTION (1937)

OIREACHTAS	GOVERNMENT	JUDICIARY
– President (1)	(15 members)	
– Dail (166 members) (2)		
– Seanad (60 members) (3)		

FUNCTIONAL	CENTRAL	TERRITORIAL
(State-Sponsored Bodies)	(Civil Service)	

COMMERCIAL	NON-COMMERCIAL	DEPARTMENTS	AGENCIES	REGIONAL	LOCAL
	(Public Enterprise)				
41	c.60	17	c.40	25(4)	115(6)
					107(7)

Full Time Employment 1984

| 88,000 | 5,000 | 37,000 | | 59,000(5) | 31,000 |

Total (8)   220,000

NOTES*

(1) Elected by popular vote, but contests not frequent.
(2) Elected in multi-member constituencies (3–5) by single, transferable vote.
(3) Forty-three elected by electoral college by Dail, Seanad and 31 county authorities; six elected by universities; eleven appointed by Taoiseach (Prime Minister).
(4) Regions not standardised; typically seven to eight for each service.
(5) Almost all in health service.
(6) Directly elected local authorities – 84 of them small to medium sized towns. Total number of councillors *c.* 1,500.
(7) Miscellaneous – harbour authorities; fisheries boards; game councils, etc.
(8) If one adds security forces and teachers to the total, the 'public service' comes to 298,000, or 27 per cent of the total persons at work.

SOURCES*

Administration Yearbook and Diary (1986), p. 354 and T. J. Barrington, *The Irish Administrative System* (Dublin, Institute of Public Administration, 1980), pp. 18–19.

expenditure as a ratio to national income just doubled between 1960 and 1980, rising from about 37 per cent to some 73 per cent.[16] So, government bulks large and is at the heart of society. How well has it performed?

The post-war history of Irish government falls roughly into three periods: 1945–58, a time of 'social Keynesianism'; 1969–72, a time of vigorous economic and social intervention; 1973–86, a time of troubles; with now, perhaps, the beginning of recovery.

1945–1958

The issue of more comprehensive social development, stimulated by Beveridge and the post-war reforms in Britain, was fought out in relation to health and income maintenance – state intervention in housing having long been accepted. During the 1950s the foundations of a modern health service, based on local government, and a comprehensive income support system, almost entirely based on central government, were laid. (Educational reform had to wait until the 1960s.) There was the hope, to be proved illusory, that social expenditure on capital and current account would stimulate flagging economic growth and so prove self-sustaining. In the mid-1950s, the onset of a major recession, with alarming deficits in the balance of payments, and a crisis of morale, sparked off by the re-emergence of heavy emigration, gripped the country.

1959–1972

In late 1958, the first *Programme for Economic Expansion, 1959–1963* was published, beginning an economic (and later a social) planning system. The Local Government (Planning and Development) Act, 1963, gave local authorities extensive powers for infrastructural development. The aim was to stimulate European levels of growth. Public capital investment was diverted from 'social' to 'productive' purposes. With varying levels of success the industrial and agricultural sectors were developed and modernised. There was a major emphasis on export-led growth. The country was being prepared for membership of the European Economic Community. The planning system had its ups and downs, but economic growth surged ahead, averaging 4 per cent a year under the First Programme, a momentum that was, despite some variation, to be maintained under its two successors. One consequence of the new affluence was a good deal of social unrest, focusing particularly on pay settlements. Wages rose rapidly. From the mid-1960s there were major improvements in income support levels, in the education system, and in the health services. One aim of social development was to raise the social services, *mutatis mutandis*, to the UK levels operating in Northern Ireland.

Many emigrants of the 1950s returned with their young families. Marriage and birth rates rose. For the first time in almost a century the population began to grow, rapidly. Self-confidence rose too high in the late 1960s and early 1970s and various corporate groups began to trample their way to the trough. Amid these particularistic preoccupations the economy began to falter, as did the administration itself, an assessment illustrated by both the abandonment in 1971 of the *Third Programme for Economic and Social Development, 1968–72*, and the failure for over a decade to produce a credible successor.

1973–1986

The Northern Ireland troubles led, from 1969, to continuing political and financial burdens exacerbated by the energy crisis and the beginning of world recession. There was also the upheaval – mainly beneficial – caused by EEC membership after 1973. There was a steady rise of exports and growth in hi-tech industry. Over-confidence led to recklessness. It was calculated that a small weak economy could spend its way out of its difficulties and, because its credit was good thanks to past prudent management, the government now borrowed prodigally at home and abroad. Since 1972 the Budget has not been balanced on current account and, notwithstanding recent serious efforts, the deficit in 1985 was 8.2 per cent of GNP. As a proportion of GNP, capital investment – the substantial part of it public investment – was among the highest of the 24 countries of the OECD and in 1979 it was the highest, at 39 per cent.

The various corporate groups, particularly public service groups, pressed their claims remorselessly. Inflation took off; between 1973 and 1983 it fluctuated between 10 and 20 per cent per annum. Excessive costs played havoc with much of traditional industry. The costs of labour-intensive public services, such as health and education, rose dramatically; each now costs about 7 per cent of GNP notwithstanding recent efforts at control. Welfare expenditures became the greatest part of rapidly rising current public expenditure. Much of the high public capital investment proved wasteful. The physical planning system, in which high hopes had been placed, is in disarray, as witnessed by a recent assessment of planning in the capital city revealingly titled *The Destruction of Dublin*.[17] Many public services have declined in quality. The inadequate road system began to break up and the telecommunication system to break down. Crime rose sharply, and police, courts and prison systems seemed unable to cope. The central taxation system – especially the taxation of incomes – is barely holding together, with the tax-paying public in sullen revolt and government, notwithstanding sophisticated and radical advice from a Commission on Taxation (1979–1985), scared to take its finger from the dyke. A stagnant labour market cannot absorb the numbers of educated young people and unemployment, at over 17 per cent, is now some 50 per cent above the EEC average.[18]

By 1986, after nearly four years in office, the government was raising the quality of public investment. The growth of current expenditure has been restrained but deficits continue. Reforms (discussed below) of the civil service and local government were promised, and minuscule taxation reforms were introduced. Pay pauses have had some effect, and inflation was falling to some 5 per cent and heading downwards.

Apart, therefore, from the decade of the 1960s, the post-war record of government is not good. A pungent critic, Professor J. J. Lee, argues that one can say the same of the whole period of Independence:

> We cannot avoid the conclusion that we have incomparably the worst record since 1921 of any economy in Northern Europe, except the British ... We are now perched, through our own efforts, at the wrong end of virtually every league table ... The first prerequisite for improvement

must be to recognise the dimensions of the problem and to admit just how deeply disappointing has been our economic performance, not just in the recent splurge of collective inanity, but for far longer. Even the growth of the sixties, however impressive by our own previous performance, still fell well below the European average.[19]

He points to the culprits:

I am deliberately ... arguing on the chosen ground of the centralisers. By their own criterion of efficiency, they have failed. Indeed, the more one contemplates the comparative trajectories of European performance, the more meretricious does the claim of centralised efficiency appear.[20]

How centralised *is* the state? By a number of criteria the degree of centralisation is very great by European standards, especially by north European standards. Irish central government revenue is an above average proportion of GDP and an exceptionally high proportion of total taxation.[21] The Dublin conurbation is expected, on recent trends, to have 40 per cent of the state's population by the year 2000[22] – the average European metropolitan proportion is usually less than 20 per cent. Dublin also contains a 'centralisation of intellectual resources unique in Northern Europe'.[23] The dynamic service sector is heavily concentrated in Dublin and the growth of white-collar employment, because of the preponderant influence of government, direct and indirect, has been largely concentrated in the city.[24] The last word – dissatisfaction devoid of nuance – must lie with the electorate: since 1972 there have been six elections, at each of which the outgoing government has been sacked. Excessive centralisation is not the whole cause of poor government but it lies at the root of many weaknesses that, collectively, lead to failure.

REFORMS?

Concern about the system of government spilled over into thinking about reform. The problems of growth – congestion, fragmentation, and declining coherence and efficiency of government – bred sporadic enthusiasms for 'decentralisation', regionalism, civil service reform, and local government reform. But thinking did not lead to action, and in 20 years little was achieved.

The first enthusiasm was for 'decentralisation', but the term referred to something far removed from genuine decentralisation. In effect, the policy was to slow the growth of Dublin by dispersing blocks of routine civil service work to new civil service offices in provincial centres; a distribution of 'goodies' which added a little more fragmentation while maintaining centralisation of power. Centralisation and Dublin continue to grow.

During the 1960s there arose an enthusiasm for regionalism and regional structures, because larger units were deemed more 'efficient' than the historic counties. A miscellany of single-purpose regions of, typically, three or four countries, was devised, notably for tourism, physical planning and health. The larger ministries and state-sponsored bodies also had, for their own management purposes, regional structures (as well as, beneath them, county and sub-county structures). The regions were not standardised in area, in degrees of discretion, or, indeed, in any other way, so they normally related, not to one another, but to Dublin. The potentialities of ordered 'regionalism' for co-ordinating public services and for injecting horizontal inputs were either not grasped or rejected. There was a horror of anything that savoured of regional 'government'.

One purpose of the European Economic Community was to encourage 'convergence' of its economies, in part by the regional development policy. The West of Ireland, along with Southern Italy, was the poorest part of the (then) Community of the Nine. Consequently, Irish regionalism looked to have some future until the Irish government, true to form, got the whole state classed as a single region!

Regionalism was killed politically by experience with the major regions, the health boards. The county authorities in a region nominated a majority of board members. Apart from three nominees of the Minister for Health, the rest were elected by the health professions. In the 1970s the local authorities were relieved of any contribution to the funding of the health services, in effect delivering the councillors into the hands of the corporate interests. If 'regionalism' meant being reduced to such impotence, then councillors wanted no more of it!

Disenchantment now seems to have spread to the centre. Health boards with no direct financial responsibilities were reluctant to take responsibility in hard times for cutbacks to combat rising costs. If current official 'leaks' are to be credited, the eight health boards are to be reduced to three, if not one. 'Regionalism' emerges in its true colours − a staging post on the road from territory to function; that is, to centralisation.

Civil service reform represents the third enthusiasm. In 1966 the problems of the overall management of the administrative system, especially the role and policy formulation tasks of the civil service, caused some public discussion. A commission − the Public Services Organisation Review Group (Devlin) − was set up to examine the organisation of the Departments of State at the higher levels and the allocation of functions between both them and other bodies. The Devlin Group reported in 1969, concluding that better policy formulation and implementation required defined and systematic roles for administrative institutions. It argued that the role of the civil service should be to review and formulate policy; that executive activities should be systematically delegated to executive agencies similar to state-sponsored bodies; that local authorities be given comparable ranges of discretion; and that the whole be held together by sophisticated planning and financial networks and a more integrated public service. A special Ministry of the Public Service was set up in the wake of the report but, in practice, little happened.

Concern about the quality of the civil service would not go away, however, and in September 1985, after a difficult gestation, a White Paper, *Serving the Country Better*, was published − the first of a promised series. It is largely devoted to improving day-to-day management practices, with the stress laid on financial and personnel management and the need to give the civil service a more human face. The most important change is the adoption of the Devlin proposal to replace, for executive tasks, the legal fiction of the omni-competent minister with executive offices (or nominated individual officers) acting on their own responsibility. Admirable as far as they go, there must be doubts about how effectively the proposals will be implemented. Shortly afterwards the Minister responsible for these reforms became Minister for the Environment with the task of modernising local government; the spectre of reform seemed to be acquiring substance.

Finally, there was the enthusiasm for local government reform. In May 1985 a 'policy statement' on *The Reform of Local Government* had been published. The major forces pressing for reform were said to be: social change, urbanisation and the growth of the capital city; the growth in traffic, recreation and tourism; the consequential extension of local government functions and the breaching of town boundaries; the feeling of irrelevance in and about local government; the growing number of voluntary bodies and their varied and variable relationship with the system; and the need for that system to be more relevant, accessible, responsive and efficient. The reform programme included the widening of local authority competence and discretion; devolving central services; some structural changes, especially in the Dublin area; minimal electoral reform in the interest of reducing voting inequality; financial reform; and, possibly, the removal of central controls. The Policy Statement was short on specifics but explicit that the 'legislation needed will be published before the end of 1985' (s. 2.3). However, by late 1986 nothing had appeared.

Even on the less difficult issues of electoral reform, voluntary bodies and boundary changes for the Dublin region, the specific proposals failed to live up to either expectations or the reality of the problems. The major aims — discretion, devolution, financial reform — are highly ambitious, but the Policy Statement does not set them in the necessary, wider context of either the governmental system as a whole or the strengthening of democracy, so crucial to the success of the whole enterprise. Promises about enlarging discretion have often appeared, but action has been in the opposite direction. The statements about devolution have been greeted in the bureaucracy, at least, with a palpable lack of enthusiasm. The question of the very existence of local taxation is a highly contentious political issue. No wonder legislation has not appeared.

There is no mystery surrounding what did appear. In November 1985 the Minister for Education produced a Green Paper, *Partners in Education*. Part of the secondary level of education is handled by 38 Vocational Education Committees (VECs), technically committees of the larger local authorities but legally separate. The Green Paper's stated objectives are to: rationalise and coordinate educational administrative structures; make them more decentralised, more democratic and more flexible; devolve to local bodies some welfare and developmental functions; achieve some degree of community orientation; and to act consistently with the Government's stated policy on local government reforms.

The actual proposals, however, would remove all secondary level education from the 38 VECs and would 'regionalise' it under 13 regions with minority VEC representation — more 'Do-It-Yourself' regionalism! In addition, some of the VECs run third-level colleges which are also to be transferred to regional-type bodies (different of course from the second-level ones) and again with limited VEC representation.

The replacement of 38 local authorities by 13 nominated bodies would strike most people as centralisation, as the reverse of devolution, as a movement away from democracy, and as inconsistent with government policy on local government reform. But for connoisseurs of Newspeak, Irish style, this sentence from the first paragraph of the Green Paper ought to be recorded (my emphasis): 'A *regionalised* education service is *consistent* with the Government's stated policy in regard to local government reform which embraces the principle of *devolution* to *local bodies* of matters affecting the welfare and development of the *communities* which they represent'. Irish centripetalism is a formidable force.

Governments may propose but self-defending public institutions dispose, and one is left wondering whether there is a fully responsible centre to this intensely centralised society. Functionalism is not only a product of centralisation but also the source of its disintegration. It does not eliminate cleavage and conflict of interest: it intensifies them at the heart of the system. Functional and institutional forces recruit ministerial voices to fight their cause around the Cabinet table itself. When these forces are powerful and self-cancelling the casualty is the coherence, consistency and overall responsibility of government. Responsibility, as it is hoarded, withers away. One argument for widely based democracy is that to divide responsibility is to propagate it, to create a territorial counterweight to functionalism and corporatism.

Irish society is moving rapidly away from its nineteenth-century origins and essentially, towards the norms of late twentieth-century Europe, an adaptation that experience suggests will take some distinctive forms. Its public institutions are locked into ideas deep frozen since early in this century. They are apparently incapable of learning from the experience. They monopolise access to the exercise of responsibility both at the periphery and at the centre itself. The nemesis of centripetalism may yet prove to be collapse into the political equivalent of a 'black hole'.

NOTES

1. M. Moynihan, *Speeches and Statements by Eamon de Valera, 1917–1973* (Dublin and New York: Gill & Macmillan, 1980), p. 466.
2. M. Fogarty, L. Ryan and J. Lee, *Irish Values and Attitudes: Irish Report of European Value Systems Study* (Dublin: Dominican Publications, 1984).
3. J.S. Donnelly, *The Land and People of 19th Century Cork* (London: Routledge, 1975).
4. S. O'Faolain, *King of the Beggars* (Dublin: Allen Figgis, 1970).
5. M. Friedman and C. Weaver, *Territory and Function: The Evolution of Regional Planning* (London: Edward Arnold, 1979).
6. A. Mathews, 'The Economic Consequences of EEC Membership' in D. Coombes (ed): *Ireland and the European Communities* (Dublin: Gill & Macmillan, 1983), p. 118.
7. Ayd [Administration Yearbook & Diary] (Dublin: Institute of Public Administration, 1986), p. 359.
8. Commission of the European Community, *The Regions of Europe*, Second Report (Strasbourg: European Communities, April 1984).
9. D. Fennell (ed.), *The Changing Face of Catholic Ireland* (London, Dublin, Melbourne: Chapman, 1968).
10. J.H. Whyte, *Church and State in Modern Ireland, 1923–1979*, 2nd ed. (Dublin: Gill & Macmillan, 1980); R. Barrington, *Health, Medicine and Politics in Ireland, 1900–1970* (Dublin: Institute of Public Administration, 1987).

11. NESC (National Economic and Social Council) Report No. 51: *Personal Incomes by Region* (Dublin: Stationery Office, 1980).
12. Fennell, *Beyond Nationalism, The Struggle against Provinciality in the Modern World* (Dublin: Ward River, 1985).
13. New Ireland Forum, *The Cost of Violence Arising from the Northern Ireland Crisis* (Dublin: Stationery Office, 1984).
14. Devlin Report, *Report of Public Services Organisation Review Group, 1966–69* (Dublin: Stationery Office, 1969).
15. P. O'Halpin, *The Chief Executive in State Enterprise* (Dublin: Irish Productivity Centre, 1979); NESC Report No. 49, *Enterprise in the Public Sector* (Dublin: Stationery Office, 1980).
16. Dept. of Finance, *A Better Way to Plan the Nation's Finances* (Dublin: Stationery Office, 1981).
17. F. McDonald, *The Destruction of Dublin* (Dublin: Gill & Macmillan, 1985).
18. F. Litton (ed.), *Unequal Achievement: The Irish Experience, 1957–82* (Dublin: Institute of Public Administration, 1982).
19. J. J. Lee, 'Centralisation and Community', in Lee (ed.), *Ireland: Towards a Sense of Place* (Cork: University College, 1985), p. 88.
20. Lee, p. 96.
21. OECD, *Revenue Statistics of OECD Member Countries, 1965–1985* (Paris: OECD, 1985).
22. AFF (An Foras Forbartha), *Ireland in the Year 2000* (Dublin: An Foras Forbartha, 1981).
23. Lee, p. 85.
24. NESC Report No. 28: *Service-type Employment and Regional Development* (Dublin: Stationery Office, 1977).

The West European State: The Territorial Dimension

L.J. Sharpe

This contribution is not about a particular country, but, rather, is intended as a synoptic overview of a central theme of the collection. It has two main objectives, the first being to emphasise and illustrate the tendency to ignore or disguise territoriality and the effects of space on phenomena in political science despite the fact that so much of politics and government tends to be inherently tied to territoriality. Secondly, it attempts to bring out some current aspects of territoriality in the political process that seem to have a special relevance to the collection.

THE REDISCOVERY OF TERRITORIALITY

Until fairly recently the concept of territoriality did not usually attract much interest in political science, perhaps even the social sciences generally, except tangentially, or in the case of political science where it is unavoidable given the subject matter, as in the study of national boundary conflict, local government or federalism. Three notable exceptions to this rule are Fesler's classic study of administration,[1] the collection edited by Maass[2] and Williams' urban study.[3] Territorial, spatial or even scale effects were usually left to the geographers. The reasons seem to be partly because the dominant modes of analysis were usually sociologically inclined and were thus cast in non-spatial horizontal categories such as class or social strata; or in vertical, sectoral categories related to the division of labour. Alternatively, the emphasis has been on the underlying primordiality of the economy, but an economy rarely defined territorially. Social or economic forces were assumed to operate in a kind of non-spatial limbo.

The other unit of analysis much favoured in political science has been, of course, the individual. But, again, the impact of territoriality on behaviour except for nationality has not been a prominent feature of the literature, and the individual is usually assumed to act in a non-spatial context. The preferred form for examining individual behaviour has been in terms of personal characteristics: party allegiance, occupation, class, religion, age and gender. This anti-territorial individualisation of political relationships has not been confined to the academic world. A major assumption of the politics of the post-war democratic state has been to view as the only dimension of inequality what Barry has called 'universalistic' egalitarianism, which he defines as 'if no individual is discriminated against then no groups can be'. This version of egalitarianism, he argues, has prevailed at the expense of 'communalistic' egalitarianism, i.e. equality between territorially distributed groups.[4] Perhaps it was a reaction to this tendency that lead to the upsurge of territorial politics.

At one level there is perhaps no mystery about the tendency of political science to sup territorial pottage with a long spoon: the recognition of geography implies uniqueness rather than universality. There are Catholics, semi-skilled white-collar workers, or socialist parties in a number of countries. They can, therefore, be treated as uniform categories irrespective of domicile. Comparison is possible and thus political science is possible. No political scientist can fail to cast an envious eye at the theoretical sophistication of economics in which variation based on territoriality seems to have disappeared. Territoriality may also have been frowned on in some quarters, especially after the Second World War, because of its association with geo-political theorising about heartlands and imperial or racial destinies.

Today, the situation is somewhat different, and territoriality might even be said to have become fashionable. Certainly at the intra-state level there has been a perceptible increase in emphasis. Space forbids an exhaustive explanation, but a reaction to the dominance of universalistic egalitarianism just noted could be one source of resurgence. Rokkan and Urwin identify three reasons; first, there has been increasing alarm throughout the older industrialised states about industrial decay. Secondly, there has been the growth of state activity in regionalised economic planning. Finally, Rokkan and Urwin note a growing unease about the centralised character of the welfare state.[5]

The state itself has also been 're-discovered'[6] and although some of its crucial territorial dimensions still remain neglected, one important aspect of the state's rediscovery as a unit of political analysis has been an interest in spatial effects,[7] both external and internal to it, not least the concept of centre and periphery.[8] A recognition that the relationship between political entities may be affected by their juxtaposition in space is also apparent in dependency and internal colonialism studies. Although now a flourishing subtopic, care must be taken not to inflate this aspect of the territorial trend, for, in some cases, the new focus is more apparent in the wording of the label; the wine in the bottle remains very much the same. In short, 'centre and periphery' and 'territoriality' have become to some extent vogue words that are thought to be more acceptable than the conventional 'central–local relations' or 'regional politics', in much the same way that 'policy studies' has often replaced 'public administration'. This claim is not, however, applicable to Stein Rokkan who is perhaps the father of the centre and periphery as a conceptual framework for getting to grips with the origins of the modern democratic state.[9] Other commendable studies of territoriality at the intra-state level include those of Tarrow.[10]

One change in the real world that has certainly been an important source of a new political science interest in territoriality was the upsurge in regional ethnic nationalism in many of the multi-cultural advanced industrial democracies in the 1960s and 1970s and the institutional responses by such states to the phenomenon.[11] To some extent this change in the real world seems to have partly influenced a renewed academic interest in what may be called macro-territoriality, that is to say in the evolution of world systems of states.[12]

These various developments apart, it is fair to say that the territorial dimension of politics still tends to be under-explored despite the undoubted centrality of territoriality for any definition of a political jurisdiction; as Gottman states, 'A political theory that ignores the characteristics and differentiation of geographical space operates in a vacuum'.[13] All polities, whether nation states, constituent states of a federation, or local governments, have to have a spatial dimension − a jurisdiction − and it is this necessity which distinguishes them from all other forms of political organisation. Max Weber, who as a sociologist was unlikely to overstate the importance of spatial factors, saw territoriality as one of the three basic distinguishing characteristics of the state which he defined as 'a compulsory association with a territorial base'.[14]

It is the boundary of the polity which also defines the limits of the sense of shared identity that gives the polity its logic. That is to say, the polity is defined by the limits of a willingness among the population to accept its laws, to be willing to be taxed by it, and in the case of the state to take up arms in its defence is necessary. In a non-democratic polity such shared identity may be problematic, since the necessary co-operation and obedience can be enforced. In a democratic polity, however, although adherence to the law may at times have to be enforced and an instilled acceptance of sacrifice on behalf of the collectivity may linger from a non-democratic past, sooner or later if the democratic polity is to remain a permanent organisation, its territorial extent must define with reasonable accuracy the limits of an unenforced sense of shared identity. This does not necessarily imply a continuously expressed sense of loyalty, or indeed a deeply felt sense of community, but it does require that those citizens living within the boundary see themselves as having more in common with one another than they do with people living outside the boundary. It is this sense of difference that makes possible the minimal conditions of an effective democracy, namely, the uncoerced willingness of the citizenry to accept the burdens that the polity may impose, irrespective of the benefits that the polity may confer.

If territoriality is one of the fundamental characteristics of any polity it is critical for the democratic polity. It also has other implications especially in relation to scale. For example, the 20 or so states that are usually included in the 'advanced industrial democracies' (AID) group vary very widely in physical scale, population size and density, with two of the extremes being the United States with 260 million inhabitants in an area of $9\frac{1}{3}$ million square kilometres, and Luxembourg with a million inhabitants in an area of $2\frac{1}{2}$ thousand square kilometres. Such disparity raises the question as to whether the category of AID states is adequate. Is the fact of a shared form of economic organisation, governmental form and concomitant social order so overwhelmingly salient as to blot out completely territorial-cum-scale effects? Perhaps. But the question is very rarely even put. The important exceptions include Dahl and Tufte[15] and the literature that sees small states as having special democratic advantages over large states,[16] or as having a better capacity for dealing with heterogeneity and conflict,[17] or, finally, that claims that small states are better at coping, both in external and internal terms, with the exigencies of the modern world market.[18]

Scale alone is clearly not a sure guide, especially for democratic stability, as the case of Western Europe amply illustrates. Belgium, one of the smallest

AID states, has been racked by internal dissension for the past 20 years. Northern Ireland is another example of propinquity diminishing rather than enhancing that underlying sense of common identity among the population that is the bedrock of the modern democratic state.

Nor can we assume, given modern forms of communication, that physical scale is an automatic barrier to the maintenance of some form of cohesive political community. Above all, if large scale is a handicap it can be moderated by territorial sub-division both in a functional sense and in terms of political allegiance. That is the efficient secret of the federal principle and perhaps the local government principle as well. The three physically largest Western states – Canada, the United States and Australia – are all federations. Nevertheless even in federations there must remain some critical questions related to scale and territoriality. Does an American living in Hawaii, for instance, some seven thousand miles or so from Washington, have a different conception of the meaning of representative democracy from a Luxembourgeois with his seat of government almost in his lap, so to speak, and his whole country discernible from a moderately sized hill? Does scale also have a more fundamental bearing on key characteristics of the state? For example, does the long-standing American predilection for the market over government have something to do with the sheer scale of the United States which makes the market feasible to some collective consumption services to an extent that would be ruled out in a smaller democracy? Equally, the tendency for federal states to have a smaller public sector may have as much to do with territorial politics, that is to say the constituent units' hostility to central expansion, as it has with any right-wing anti-statist ideology.

Even when scale seems to be formally recognised institutionally as in a federation, there may be a high price to be paid in terms of 'democratic-ness'. Also federalism may be a two-edged sword if the boundaries of one of the constituent states encapsulates a minority culture which feels aggrieved, as the example of Quebec in Canada suggests, at least when it was controlled by the *Parti Québecois*. Far from assuaging minority aspirations by matching its desire for recognition of its sense of difference by some form of autonomy, federalism may feed such aspirations to the point of demanding outright separation. The long line of failed federations since 1945 remind us that federalism is not in any sense a panacea for uniting the un-unitable. The various attempts in South America, the West Indies, Central Africa, East Africa, Mali, Ethiopia and Eritrea, Malaysia, Pakistan and the five failed attempts to create an Arab federation in the Eastern Mediterranean all testify to this proposition. Moreover, federalism, in a Western European context, must always involve a trade-off between democracy and constituent state equality. Thus the greater the emphasis on the federal principle the greater the sacrifice of the largest constituent units.

The reality of territoriality does not, of course, confront only federal states; all democracies have to cope with it by some form of institutional arrangement to the extent that even the very smallest do not seem to have been able to dispense with local government. On the contrary, some evidence suggests that sub-national government in general in most Western states has tended to flourish since the end of the Second World War.[19] Indeed, one commentator has been led to conclude that such is the prevalence of territorially shared power in *all* modern states that the time-honoured category of the unitary state is now redundant.[20]

Viewed in this way, it is difficult to escape the conclusion that territoriality is a fundamental part of the study of the democratic polity. Yet, too many major territorial aspects of government and politics still await systematic study. It is surprising that we still await a comparative study of the 'red capital' phenomenon — Vienna, Berlin and London, not merely as examples of big-city politics, but their role in national politics as well. Nor has anyone tackled the similar phenomenon of the 'red island' — comparatively — that is to say, the radical city in a sea of conservative ruralism with which it seems to have a symbiotic relationship. Equally ignored is the intriguing operation of what may be called unobtrusive federalism, whereby the heavily rural backward half of a state compensates for its economic inequality by penetrating its political institutions. The *Mezzogiorno* in pre-Fascist Italy is one example of this phenomenon as is Occitania in the Third Republic of France. So, too, is the South of the US which created a kind of 'double federalism' from the inception of the Republic up to the 1960s with the brief interregnum of the Civil War and Reconstruction.

These are fairly specific gaps, but to shift to more formal terms there seem to be four senses in which the importance of territoriality is most evident, three of which tend to be seriously neglected.

THE BOUNDARY DETERMINES MAJORITIES

The first aspect of territoriality that is fundamental to understanding the modern democratic polity and, it must be conceded, the least neglected, is the fact that the boundary of a polity, its physical extent and shape, may critically determine who is and who is not included and thus the composition of the majority. It is an aspect of territoriality that is least neglected for the obvious reason that it may involve the integrity of the state itself and is thus at the heart of most international conflicts. At a less exalted level, this aspect of territoriality may feature, for example, in the problem of revising, or not revising, area-based electoral systems where there are large changes in population distribution. This dimension of territoriality is thus a thriving industry in all those states stemming from the British tradition of area-based constituencies for election to national representative assemblies. The territorial determination of majorities may also intrude into the reorganisation of local government boundaries. Certainly this is the case in the UK where there can be no doubt that every phase of the post-war reorganisation process has been influenced by estimates of the likely effect of boundary change on the fortunes of the parties.[21]

THE UNIQUENESS OF THE TERRITORIAL ENTITY

The second and more neglected political dimension of territoriality is simply the fact that each polity — nation state, constituent federal province, or local government — is unique in the sense that its location in space cannot be shared by any other polity. This uniqueness means that it will be faced with some inescapable tasks that are not reducible to some feature of its social or economic structure, stage of development, relative wealth etc., but from the advantages and disadvantages that it enjoys because of its location in relation to other polities, and to any other spatially related characteristics like climate and physical context (island, landlocked, close to major trade routes and so forth). In short, all polities are engaged in activity that is territorial in origin and is purely political in the sense that is is unique to the polity and not an epiphenomenon of its internal social or economic character. No one could, for instance, begin to understand the long, bloody history of Flanders without recognising the fact that it lies athwart the frontier dividing the two most important linguistic and cultural groupings in Western Europe that have remained largely unchanged since the fifth century – the Latin and the Teutonic. Similarly, no one is surprised that Finland does not take part as, say, the Netherlands does in NATO or the EEC despite the fact that it indubitably forms part of Western Europe in ideological and cultural terms. It is the recognition of this aspect of territoriality – that states have functions derived from their location in a wider system – which inspired some of the earliest centre and periphery analysis by Stein Rokkan, particularly his conception of the city-state chain strung along the great North–South river axis of the Rhine and the Rhone.[22] Some of Rokkan's other supra-state conceptionalisations were somewhat less persuasive, however, particularly his acceptance of the geo-economic designation of a 'European core' which, by definition, must imply some determinant external to the states themselves. It turns out, however, to be merely a line encompassing the cores of most West European states. This lends it a certain degree of plausibility, except that such a designation means that the epicentre of the economic heartland lurks somewhere among the vineyards of the Mosel valley and the mountains of Western Switzerland, and the Glasgow–Edinburgh urban concentration and Bilbao are excluded whereas the Massif Central is included.[23]

TERRITORIAL ALLEGIANCE

The third, relatively neglected political consequence of territoriality is the fact that an individual's location can determine the collective interest he shares as much as his ideological predilections, his occupation, his religion or his class. In some circumstances his commitment to a locational allegiance may override the other forms of commitment. At the level of the nation state it almost certainly will. The modern dominance of the economic and social explanatory mode in political science, and perhaps the social sciences generally, has tended to understate this likelihood. To take a mundane example, it is very rare nowadays to see patriotism seriously examined as a component of right-wing voting. At the heart of all politics, it tends to be assumed, lurks something more fundamental; it might be social, but much more likely it is economic. Here neo-classical economics joins hands with Marxism to form an extremely influential economic determinist assumption to which a great deal of political science appears to be prone. This economic determinist inclination sees political processes as largely contingent, or mere epiphenomena, and political institutions may even be seen as an encumbrance that distorts the free expression of individual choice. Neo-classical economics, for example, regards the state at least by implication, as an impediment to a rationally operating market where individuals are free to maximise their utility functions. In their terminology it is a 'discontinuity': as Douglas North has noted, in the neo-classical model there 'are no organizations or institutions except for the market'.[24]

Orthodox Marxist doctrine is equally dismissive of the state, since it is defined merely as a bourgeois construct designed to promote the interest of capital and thwart the creation of international working-class consciousness. The 'working men have no country', noted Marx and Engels. The fact that in the reality in which both doctrines find their strongest behavioural expression the state is a trifle less contingent than either theory can countenance is rarely recognised even by those who would not claim allegiance to either theory; such is the power of an economic explanation. So there is a curious disjunction whereby both theories flourish at the same time as their living refutation. To claim that the state refutes neo-classical theory may be resisted by some on the grounds that neo-classical theory does not refute the state, but merely claims it is irrational. However, there can be no market without a state of some sort. No market, world, regional or national, can possibly function without them, if only to maintain the legal order which necessarily underpins all markets. For its most important form, that within each nation state, the market is entirely dependent on a prior order since it could not operate without the enforcement of contracts.[25]

A similar dependency on the state exists in relation to Marxism, for it has found perhaps its greatest triumph not as the supra-national *Weltanschauung* of the international working class, but in the *perpetuation* of nation states. In the post-war era Marxism constitutes one of the most popular doctrines, feeding post-colonial state nationalism for which it furnishes a critical 'horizontalising' function: it transforms an essentially spatial and vertical distinction into a transnational horizontal or stratification concept. The ex-colonial state becomes the functional equivalent of an oppressed class. Its bid for self-determination and its progress (once independence is achieved) is thus given the same moral dimension that may accrue to the exploited working class in the capitalist system. This transformation is crucial, for it renders what might otherwise be regarded, especially after independence, as just one nationalism against others with no more and no less to commend it, into a relationship about which moral distinctions can be made.

This horizontalising ploy is nothing new in the history of nationalism and was critical for nineteenth-century progressive liberals who favoured pan-Germanism, Italian unification and freedom for the Greeks, but baulked at according the privilege of a language state to, say, the Croats, the Czechs or the Irish. The horizontalising ingredient then was not an egalitarian one but one based on high culture. Only those nations with an internationally recognised high culture deserved their own state.[26] It is of some interest that Engels, although by no means free of the superior–inferior culture view, chose a slightly different stratification principle and drew a distinction between 'historic' nations, such as the Poles or the Scots who had enjoyed a previous existence as separate states, and *geschichtlosen Völker* such as the Basques and the Welsh who had not.[27]

J. S. Mill had his own stratification order for nationalities which seems to have combined the progressive liberal high-versus-low culture with the decidedly harsher *geschichtlosen* doctrine of Engels. For Mill there were 'highly civilized and cultivated people' such as the French and the English who provided the opportunity for their respective 'interior and more backward' peripheral enclaves – the Bretons, Basques, Welsh or Scottish Highlanders – not to go it alone but, rather, to give up their backward ways and transform themselves into Frenchmen and Englishmen. That course, argued Mill, was much more preferable for a member of a peripheral nationality, '... than to sulk on his own rocks, the half savage relic of past times, revolving in his own little mental orbit, without participation or interest in the general movement of the world'.[28]

The point to note is that Mill's purpose is precisely the same as, say, a latter-day dependency or colonialist theory, namely to smuggle into a centre–colony conflict a rationale that made it possible to take sides on grounds that are more congenial than would be possible if the conflict were regarded simply as a struggle for power.

To return to the mainstream of our discussion, the tendency for modern social science to understate territorial determinants of political action and allegiance may also be illustrated at the level of the peripheral region, for instance, in the surprise and bafflement with which the emergence of regional cultural nationalism in some West European states was received in the early 1960s. How could it occur when almost all the predictions of an impressive array of distinguished social scientists past and present, embracing all points in the political spectrum, concerning the future of the 'advanced industrial state' had firmly indicated centralisation, homogenisation and the elimination of precisely the kind of spatially determined consciousness the new regional nationalism reflected.[29] This was, as William Beer was to put it, an 'unexpected rebellion' with a vengeance.[30] Not the least unexpected of its features was that it seemed to run counter to, and not be an epiphenomenon of, undoubtedly homogenising and centralising social and economic forces. In short, it indicates a certain autonomy for political forces that was not supposed to exist. It is, therefore, not entirely surprising that some effort was expended in rendering the 'rebellion' into a form that was more consonant with conventional social science orthodoxy. There was, thus, a number of interpretations that 'horizontalised' it in the form just discussed. The new regional nationalism thereby became a response to economic deprivation and, in the extreme cases, a response to 'internal' colonialism.[31]

It seems likely that such 'horizontalization' was, to some degree, justified, especially if we take a longer, pre-industrial, perspective. This takes into account the fact that often the dominant core culture captured the cultivable land, driving the minorities to the less hospitable and less productive uplands. In short, the internal colonial interpretation considerably enriched our understanding of centre–periphery relationships. Moreover, during the post-war period many of the cases of regional nationalism could be interpreted as arising in regions that were peripheral to a dominant 'imperial' core, and such peripheries had manifestly not always shared in the post-Second World War boom on an equal basis with that core.

The internal colonial model did not explain, however, those regional movements, often some of the most virulent, where the source of the conflict was not relative economic deprivation but arguably the reverse, such as Flanders, or the Basque country and Catalonia.[32] Nor could it explain the dramatic but successful campaign of the Francophones of the Bernese Jura for a new canton of their own. Equally, it could not account for the desire of most Ulster Catholics to join the Irish Republic – a switch that would involve a marked drop in living standards and state benefits. Still less could it explain the vehement desire for Ulster Protestants to remain in the United Kingdom where, by every conceivable economic measure, they experienced the lowest living standards of any region in the UK.[33]

The principal defect with this type of horizontalisation is that whereas it is possible to accept that there can be *some* collective economic deprivation irrespective of class or income, the theory assumes in the name of collective group inequality that significant intra-group inequalities are obliterated. Thus, the Corsicans, say, including the Ajaccio millionaire, the *département* civil servant and the indigent hill shepherd are all equally affected in economic terms by French internal colonialism. Marx is truly stood on his head, for, paradoxically, the attempt to transform a spatial entity into the equivalent of a class has perforce to abandon all notions of economic or social inequality, that is to say classes *within* the oppressed spatial entity. As Tarrow has pertinently asked: 'If the major axis of exploitation is territorial and the major villain is external, does the local bourgeoisie thus become part of the "territorial proletariat"?'[34]

Economic inequality is clearly only part of the story of that type of territorial consciousness that produces regional ethnic movements, albeit an important part. Apart from the obvious mobilising factor of a pre-existing sense of difference from the core, the upsurge of regional nationalism also seems likely to have been spurred, as we have noted, by a reaction to centralising and homogenising social and economic forces.

The desire to horizontalise sub-national self-assertion, no less than national self-assertion, into a species of class inequality, like the inferior–superior culture attempt over a century ago, reminds us that nationalism is rarely viewed with a wholly detached eye. The human spirit, it seems, even more perhaps the academic spirit, baulks at identifying all nationalisms as simply power struggles the only resolution of which may be armed conflict. Another ingredient must, it seems, be added to national conflicts which will enable the observer, if not all the participants, to transform the conflict into one based on a more congenial value such as justice, equality, culture, or whatever. One side can then be identified as having a superior cause.

TERRITORIALITY AND SCALE

The final aspect of territoriality which seems to be under-explored in political science, is the role of scale in a democracy – a problem touched on briefly earlier. Scale is as much a matter of sheer size as of territory, but the two are difficult to tackle separately. This aspect of territoriality is important for the modern democratic state because it can affect the quality of its representative system. Simply, as a matter of fact, the larger the physical size of the polity the greater the likelihood of lower access to government for citizens, whether that access is to public services or to decision-makers. On this measure, physically small polities are likely to be more democratic than larger. They may also be more democratic if they have smaller populations as well.[35]

To some extent, both deficiences of large states can be modified by federalism, as noted earlier. By dividing governmental power territorially, that is between the constituent polities making up the federation, some government is closer in distance terms to the majority. But not all public functions can be devolved to sub-national government; some public functions are indivisible, or are too expensive to devolve. It is, of course, possible for large polities to make public services, if not decision-makers, as accessible as small polities if they are willing to provide the necessary number of service delivery points. But it is highly unlikely that they will do so because the greater the physical scale of a jurisdiction, the less aware can its controllers be of territoriality – of the consequences, that is, of distance and spatial dispersion.

Jurisdictions like a polity, especially the state itself, are fixed 'givens', and decision-makers have to assume that there are no problems of popular accessibility to state services that cannot be resolved by some form of deconcentration. But it is unlikely that such deconcentration in a state such as France will achieve the degree of public access that will be available for centrally provided services in, say, Luxembourg. Problems of distance and territoriality that might question the efficacy of the jurisdiction in terms of accessibility do not arise, the existence of the state is its own justification. The federal mode of moderating the access deficiencies of large and unwieldy scale can be chosen but there may be the democratic price to be paid that has been noted, and it may be a large one. One of the consequences of the relative lack of interest in this aspect of territoriality is a tendency to ignore physical scale and configuration altogether. In the extreme case where system capacity is smuggled into the democratic formula, scale becomes a positive advantage. Where this is done the measure of democracy is not only a function of power distribution within the polity but also a function of its capacity to undertake certain tasks.[36] Such tasks are those that are thought to be determined by scale so that the bigger the state the more democratic it is because the more it can do. The ultimate – the *reductio ad absurdum?* – of this definitional transposition of democracy into a product of functional capacity rather than the distribution of power is the argument for world government which seems to assume that questions of territoriality, voter equality and access do not exist. For it is unclear how world government would operate as a democracy, let alone how the gargantuan bureaucracy it would entail is to be controlled. Yet world government is often claimed as a blow for democracy on the grounds that it has the functional capacity to end all wars. And democrats must, by definition, be against settling international conflict by force.

After this brief excursion into the inadequacy of the political science debate on territoriality, the remainder of this survey will examine certain key aspects of territoriality. The list of aspects is not exhaustive, nor does it necessarily include all the most important, but rather, focuses on those that have a special relevance to the present volume.

SOME PERIPHERIES MAY BE CENTRES

The first aspect of territoriality that merits attention is the concept of centre and periphery as a model for modern state building. In its simplest form, it defines a territorial relationship within a state between a dominant core and peripheral regions that have been absorbed by it. The core also provides the dominant cultural ethos and language for the whole state and, to a greater or lesser extent, the ethos if not always the language that has been adopted by or enforced on, the peripheral regions.

Broadly speaking, this model can be exemplified by the Ile-de-France in relation to the rest of France; Southern England and the UK; Brandenburg or East Prussia for modern Germany up to 1945; Castile for Spain; Piedmont in relation to Italy. One problem that confronts the notion of a core expanding to absorb a periphery, which has not perhaps been adequately explained by the core–periphery theorists is that it does not necessarily refer to a radial process. Nor does the core have to be like an apple core: cores may be at the geographical extremity of a state. Nor need they arise in historically-dominant centres, or where the majority of the population live: if Piedmont is the core of modern Italy, it was in historical and population terms a periphery. Brandenburg, too, was a peripheral 'core' of modern Germany; and East Prussia was even more so.[37]

Matters may be even more confusing: in some cases, as Tom Barrington's article makes clear, it is the periphery that is the guardian of the state ethos. Western Ireland – the Gaeltacht – would claim that it had a right to be seen as the birthplace of an independent Ireland equal to that of Dublin, with its strong English character, as the capital of the Pale. This reversal of the status of centre and periphery seems to be associated with states which have been previously themselves peripheries to larger states. The same phenomenon may be discerned in Norway, where Nordland, Troms and the South Western periphery would never concede that Oslo, with its strong Danish and other foreign links, had a higher claim to be the cradle of the Norwegian national spirit. A similar inversion of the periphery's role in relation to a national ethos may have occurred in Finland where Helsinki had always a strong Swedish character. Where the state itself has formerly been a peripheral region, it seems that peripheralness rather than centrality is the more valued attribute.

Sometimes, as in the case of France, as Yves Mény's essay demonstrates, the peripheries may make claims similar to those of the Gaeltacht and Nordland, but the origin of such a reversal of roles in the French case has more to do with region and political ideology than with specifically regional culture. The French case does, however, also remind us that one fundamental territorial dimension of the politics of the so-called AID states is the tension between the cities and the rural agricultural areas. The French version of this tension was merely sharper because throughout the modern era until fairly recently the Paris region was, by international standards, in the vanguard of education, technology and industry and the culture which goes with them, whereas provincial France was not. The French example also reminds us that this conflict between the industrial and the agricultural – the urban and the rural – may have a special and more intense quality in the uni-centred (or in Rokkanese, the monocephalous) states such as Denmark, the UK, Ireland and Norway, and perhaps Finland and Sweden, as well as France.

France also yields up another strangely ignored, but in the modern era critical, dimension to the rural-versus-urban form of territorial politics which has its origins in the Second World War. Wartime economies need food even more than they need the *matériel* of war, and peasants whose adhesion to the national state is likely to be of a somewhat less uncompromising quality than that of the town-dweller when the peasants' economic interests are at stake will, given sufficient incentives and a low enough morale, meet that demand for food irrespective of who makes it. Such willingness may be greater if agriculture had been in the doldrums and was discriminated against for the preceding twenty years as European agriculture was in 1940. Modern wars, in short, tend to create agricultural booms, and on the whole 'rural' areas tend to suffer on the whole less disruption and damage than cities. In 1945, then, in sharp contrast to the cities, most of West European agriculture was relatively prosperous and that prosperity was prolonged by post-war shortages of food in the ravaged cities. To put it in a nutshell, with rare exceptions, West European farmers did comparatively rather well out of the war.

In most of Western Europe the agricultural sector was, therefore, in an exceptionally advantageous position to influence at birth the form of post-war democratic politics: it was an influence greater than its numbers warranted and it was sufficiently strong to retain that influence as those numbers dwindled. From the post-war power base thus created in individual states farmers were in turn able to exert a critical influence on the formation of the EEC which even today remains still predominantly a cartel of agricultural producers and especially exporters.

Another important variant of the dominant centre and tributary periphery model occurs where peripheries see themselves not as being in any sense tributary, but as culturally superior to the core. Such a sense of superiority may occur by virtue not so much of the intrinsic merits of their own culture, but more because that culture is closer to another culture external to the state of which the periphery forms part. Thus, Catalonia's sense of superiority over the Spanish core culture is derived not simply from seeing itself as being untainted by Arab admixture and having enjoyed a longer uninterrupted period under Christianity, but also as having closer links with the heartland of West European culture to the North. Similarly, association with a wider and purportedly superior culture reinforced peripheral nationalism against the national core in the cases of German speakers of the South Tyrol in Italy.

One aspect of peripheries as centres which merits our attention is the extent to which peripheries have their own peripheries. Such a Chinese box phenomenon can, amongst other things, make the horizontalising of the core-periphery conflict into a conflict of equality described earlier even more inaccurate; for the oppressed may have to be re-defined as oppressors. For example, the Scottish National Party's highly successful progress in the late 1960s and early 1970s had the effect of heightening the sense of *non*-Scottish-ness among the Shetlanders, the inhabitants of a group of islands that were formerly Norwegian and were acquired by the Scottish Crown in the sixteenth century. Having endured enforced 'Scotification' in the past, the possibility of rule from Edinburgh did not seem to appeal to the Shetlanders. But what should be the attitude of committed Scottish nationalists to this comparable expression of a peripheral sense of injustice and of difference? As committed peripheralists how could they deny the claims of another periphery? Should the purely Scottish nationalist element in their ideology override the right to self expression element? In most cases, one can be reasonably sure that it will, for all peripheries have, perhaps have to have, a *de minimis* rule. Thus, the right to self-determination (or whatever) ceases to apply at some scale level below that of the periphery in question. One of the best examples of such a *de minimis* principle in operation occurs in the Irish Republic where the stauncher the Irish nationalist, the less he is willing to recognise that the right of self-determination that was asserted for the South when seeking independence cannot logically be denied the Protestant majority in the North. On the contrary, he is likely to assert an even more trenchant 'core' ideology against the Ulster periphery within a periphery than the British state dared to assert against the Southern Irish.

Sometimes the periphery *de minimis* rule seems to apply in some small states. For example, in the Netherlands the Frisians living in the province of Friesland number some 547,000 and constituted in the mid-1970s some 4 per cent of the total Dutch population. This proportion is broadly comparable with that of the Welsh in the UK where the population of the Principality comprises 2,730,000, or about 4.9 per cent of the total national population. The Frisians, like the Welsh, have their own written language of ancient lineage and status, the public use of which, however, until the early 1950s was under draconian legal restrictions.[38] The Frisians also have a fairly strong sense of identity which expresses itself in an independence movement of sorts, with its own cultural council (the *Frysk Akademy*) and a longstanding small party (FNP) a section of which sits as elected representatives on the provincial council, but which does not contest national elections.[39] But, whereas the Welsh-speakers, about 685,000 in number, are barely 1 per cent of the national population, Frisian speakers are estimated to be approximately 400,000,[40] which is some 3 per cent of the national population. In terms of numbers, Friesland is also relatively more prominent in the Netherlands than Brittany in France. If we take the now accepted definition of Brittany – the 600 communes of *Basse Bretagne* – this region accounts for about 2.8 per cent of the total French population and Breton speakers about 1.3 per cent.[41] The Dutch Frisians are also proportionately more important within the national state than the Carinthian Slovenes in Austria and the Bernese Francophones of Switzerland. Yet it is rare to find Friesland either as an ethnically distinctive province, let alone its political movement, even so much as mentioned in modern studies of the politics and government of the Netherlands in English.[42] An exceptionally rare example of a reference to Friesland occurs in Daalder's 1970s essay which compares the Netherlands with Switzerland, but despite the fact that such a comparison could hardly ignore language differences entirely the reference amounts to a half-sentence to the effect that Frisian is a separate language from Dutch.[43]

One obvious reason for this surprising *lacuna* is that the Netherlands is always rightly identified as the archetype of the modern pluralist state divided as it has been until fairly recently into sub-cultures, the famous *verzuiling*. Such salience must have correspondingly diminished all territorial differences within the Dutch state, including perhaps the special character of Limburg as well.

CHANGES IN THE EUROPEAN STATE SYSTEM

One of the most interesting but rarely, if ever, examined features of the European state system is that it has been slowly, very slowly, fragmenting over the past century or so; that is to say, the number of separate states has been steadily increasing since about 1875. At that date there were 20 separate European states, which is less than at any time previously or since. Today, the number has almost doubled to 36. With the exception of East Germany, the new states created since 1945 have been on the extreme periphery and most are also very small in both area and population. The only exception is Greenland, which for what it lacks in population, is more than amply made up for in area. They are, in short, examples of the world-wide phenomenon, the 'micro state'.[44]

Not only has the European state system been slowly fragmenting for a century, there has also been no state consolidation since 1945 except in the cases of Russia and Poland. The former incorporated very large tracts of Finland, Rumania and Poland, and absorbed Estonia, Latvia and Lithuania; the latter, in turn, acquired East Prussia, part of old Prussia and Silesia. Other than Russia and Poland, no other European state has succeeded in territorial aggrandisement since the end of the Second World War. So although incorporating it into its Constitution since the 1930s, Ireland has not so far succeeded in acquiring Northern Ireland; French territorial claims beginning with the Channel Islands and the Saar in the 1940s and including various threatening noises made to Monaco have all come to naught. Moreover, by 1962 France had lost the whole of Algeria which it had designated as an integral part of its national territory. Similarly, Yugoslavia also lost its bid for Trieste, and Greece failed in its attempt to absorb Cyprus despite support from many Greek Cypriots who, ironically, then lost hegemony in Cyprus as a direct result of their Enosis aspiration. Austria, although making no formal claim, has made no secret of the fact that it would like the Alto Adige back, but it is safe to say that Austria's desire is a forlorn one. Spain has yet to acquire Gibraltar despite making a number of moves, diplomatic and otherwise, designed to that end. Not only, then, has the European state system been slowly fragmenting, any further integration of existing states appears so far to be ruled out as well.

Since the end of the Second World War there has, of course, been a integrative trend, in Western Europe at least, in the form of the supra-national co-operative bodies that have been created, supreme among which are the EEC and NATO, but also include the European Parliament, the European Courts of Justice and the OECD. However, none of these co-operative arrangements is integrative to the point of eliminating existing states as sovereign national entities. The EEC has made limited inroads into the sovereignty of its member states since it has its own directly elected Parliament, can raise taxation in them and promulgate binding legislation on a small number of issues. It also has a supreme court, the decisions of which are binding on member states. To a lesser extent, NATO infringes member state sovereignty in some defence matters. If these supra-national institutions constitute an impressive and probably unique example of voluntary co-operation between sovereign states, it would be false to see such co-operation as part of some zero-sum game in which the acquisition of powers by supra-national institutions is an automatic surrender of individual state autonomy.

If it were possible to sum the total effects of these institutions in terms of the resilience of the member states *qua* independent states it would have to be a positive figure. That is to say, the growth of post-war co-operative organisations – in the case of the EEC amounting to an embryo confederation – constitutes a positive sum game. This is especially the case for the smaller member states which have gained very considerable advantages, since they have been able to participate at levels of international trade and defence that would have been closed to them acting alone. They have also had their general trading position enhanced over what it would have been had they been required to deal with the larger member states on a one-to-one basis. Above all, the smallest member states enjoy a defence umbrella of a quality they could not possibly afford if they provided their own defence. In a very real sense, the creation of West European supra-national co-operative organisations has enhanced the position of its small state members beyond measure. Since 1945, as in the world at large only more so, an order has been created in Western Europe in which the small state flourishes to an extent that perhaps has no parallel in the past.

With the probable exception of the UK, the larger states of the EEC do not seem to have sustained any loss of national self-consciousness or independent statehood either. On the contrary, both Italy and Germany emerged from the ashes of defeat as full members of the international community with remarkable rapidity, a process that was undoubtedly accelerated by membership of the EEC, and it would be difficult to name another period in the modern era when France stood higher as an independent national state.

In short, co-operative supra-national organisations, even such highly developed ones as the EEC and NATO, whatever their final form in the future, are at this stage in their evolution as much the promoters of statehood as they are underminers of it. It would, therefore, be a mistake to confuse the loss of autonomy that membership entails with the creation of new states, the territorial aggrandisement of existing states or their disappearance by annexation. Moreover, in so far as the growth of supra-national organisations has affected the stability of the West European state system it has, if anything, enhanced it. In the post-war world of inter-state co-operation, it seems, territorial aggrandisement is not acceptable. In addition, supra-national co-operation may have halted not only further state integration, it may have enhanced fragmentation; Denmark's membership of the EEC, for example, accelerated the final separation of Greenland and the Faroes from Danish aegis.

NATIONAL BOUNDARIES ARE SACROSANCT BUT NOT SUBNATIONAL BOUNDARIES

If the European state system of territorial boundaries seems to have congealed into a pattern in which the only change likely is that of peripheral fragmentation the same cannot be said of intra-state boundaries; that is to say the boundaries of the subnational government systems of European states. Within the subnational category, the boundaries of the constituent states of a federal system are very difficult to change, although as noted earlier, pressure from French speakers of the Bernese Jura created a new canton. This, however, was an exception, and the boundaries of the constituent polities of a federation are usually almost as sacrosanct as national state boundaries and certainty a great deal more immutable than local government boundaries. Over the past 25 years most West European states have been changing the boundaries of their local government systems, some like Sweden and the UK drastically, and such fundamental change leads us to an important aspect of territoriality; namely, that whatever the abstract deficiencies of a state's boundary in terms of its capacity to act as a state they have to be accepted. That is to say, there can be no question of designing national boundaries around some desideratum of functionality. Although a great deal of ink has been expended on whether certain countries are 'viable', in the last analysis states exist because their citizens want them to exist and neighbouring states respect that wish. This is not the case for a system of local government, the boundaries of which do not demarcate a national sovereignty and so can usually be changed to suit specified functional needs or changed population patterns.

The changes in the local government system of Western European countries have, broadly speaking, been made in response to the twin pressures of urbanisation, which had rendered urban boundaries obsolete, and the need to raise the quality of local services and the efficiency with which they are delivered. In general, the consequence has been that the population, physical scale and resources of local government units were increased. The most extreme case is the United Kingdom which now has some of the largest basic units of local government in Western Europe and possibly the whole of the Western world.

In what may be called the Napoleonic group of states (Belgium, France, Italy, Portugal and Spain) modernisation of the existing structure of local government has proved difficult, compared with the northern European group. This seems to be partly because under the Napoleonic model the existence of central field services and the prefect provides a strong bureaucratic interest in the status quo. The very functional inadequacy of local government that provides one of the main reasons for modernisation also provides the central field service technocracy with its *raison d'être*. Moreover, because many local government units are largely dependent on the central field services they are absolved from any inhibitions in resisting change on the grounds of functional inadequacy.

Resistance to structural modernisation will be all the more successful in Napoleonic systems in which local government has had to evolve techniques for by-passing the prefect and colonising the centre. Such techniques are in the French case pre-eminently the *cumul des mandats* tradition, and in the Italian case the party, and both make it possible for the localities to nip centred reorganisation proposals in the bud.

Faced with such formidable obstacles, the Napoleonic states have avoided fundamental changes in the local government structure, favouring instead the insertion of a new tier of sub-national government 'above' local government at the level of the region. All five Napoleonic states have created such a regional tier, or seem to be in the process of doing so, with Portugal at what might be called the embryo stage. Italy, as Robert Leonardi *et al*'s contribution to this volume reveals, is at the most mature stage, having begun the process in the late 1940s and implemented it in the mid-1970s. With a regional tier in place, many of the deficiencies of scale that we noted prompted the restructuring of the local government system proper disappeared, since all the regions are larger both territorially and in population terms than existing units of local government.

It would be idle to claim that these regional governments are solely the result of attempting to achieve the aim of local government modernisation by other means: other factors such as regional ethnic nationalism have sometimes been far more important, especially in Belgium but also in Spain. A residual echo from the late 1960s of regional economic planning may also have played a part. Nevertheless, the striking difference between the Napoleonic and non-Napoleonic groups in terms of local government reorganisation suggests that factors common to each type have been important and may have been decisive.

A parallel development to local government modernisation in the territorial politics of West European states has been a conscious attempt to decentralise state power. This decentralising trend has taken different forms in each country and has sometimes, as in Denmark and France, been combined with the modernisation of the local government structure. As Hesse's and Toonen's essays in this volume make clear, both West Germany and the Netherlands have also sometimes pursued consciously decentralist policies aimed at transferring power to lower levels of government. These stated intentions must be treated with some caution, since decentralisation now seems to be a political fashion throughout the West. Caution, if not outright scepticism, would certainly seem to be the appropriate attitude to trends in the UK where, as Rhodes' essay amply demonstrates, the trend over the past decade or more has been one of fairly intense centralisation with few pretensions to decentralisation. Even in France (where decentralist declarations were accompanied by what seemed like a root and branch transformation of *tutelle* and the creation of regional bodies that were independent of the centre's field bureaucracy) reality so far, as Yves Mény suggests above, seems much more like the *status quo ante*.

Yet, if expenditure is a measure of power then there can be no doubt that there has been a consistent decentralising tendency among the advanced industrial democracies that predates the onset of consciously decentralising policies among them. There seems to be something inherent in the character of the modern industrial state which entails a steady shift in the weight of the public sector toward the sub-national level or levels. This tendency is clearly illustrated by the fact that the share of total public expenditure spent by sub-national government was steadily rising from 1950 to 1972 in almost all the 22 Western states; in only two (Ireland and Switzerland) was this not the case, and by 1973 in over half of the 22 sub-national government was spending more than the centre.[45] There is some evidence that this trend has continued since 1973 at least in 14 AID states where, if central transfer are excluded, over the period 1960–82 sub national expenditures have consistently continued to rise at a faster annual rate than central expenditures for all 14.[46] Even in the UK, despite the best efforts of the centre, local government expenditure increased at an annual rate in excess of the centre's by 0.1 per cent. When central transfers are included for the 14 sub-national expenditure still grew at a faster rate than those of the centre in 6 of the 14 over the period.[47]

The precise reasons for this highly significant long-term shift in the balance between centre and sub-national government are almost certainly complex and multi-faceted and this is not the place to provide an explanation. But it does remind us that the periphery is not dead nor even dying; on the contrary it is very much alive and kicking, and vigorously so in most Western European states.

NOTES

1. James Fesler, *Area and Administration* (Tuscaloosa: University of Alabama, 1949).
2. A. Maass (ed.), *Area and Power* (Glencoe: Free Press, 1959).
3. Oliver P. Williams, *Metropolitan Political Analysis* (New York: Free Press, 1971).
4. Brian Barry, 'Reflections on Conflict', *Sociology*, Vol. VI (1971), No. 1, p. 5: See also Jean Gottman, 'The Evolution of the Concept of Territory', *Social Science Information*, Vol. XIV (1975), Nos. 3–4, p. 42.
5. Stein Rokkan and Derek Urwin (eds.), *The Politics of Territorial Identity* (London: Sage, 1982), p. 191.
6. Peter B. Evans, Dietrich Rueschemeyer and Theda Skocpol (eds.), *Bringing the State Back In* (Cambridge: Cambridge University Press, 1985).
7. See, for example, S. E. Finer, 'State-building, State Boundaries and Border Control', *Social Sciences Information*, Vol. 13 (1974); S. G. Tarrow, Peter T. Katzenstein and L. Graziano (eds.), *Territorial Politics in Industrial Nations* (New York: Praeger, 1978); Rokkan and Urwin; Tarrow, *Between Center and Periphery, Grassroots Politicians in Italy and France* (New Haven: Yale University Press, 1977); Charles Tilly (ed.), *The Formation of National States in Western Europe* (Princeton: Princeton University Press, 1975).
8. See, for example, Gottman (ed.), *Center and Periphery: Spatial Variations in Politics* (Beverley Hills: Sage, 1980).
9. See, for example, S. M. Eisenstadt and Rokkan (eds.), *Building States and Nations* (Beverley Hills: Sage, 1973), 2 vols; and Rokkan and Urwin; Rokkan, 'Geography, Religion and Social Class: Cross Cultural Cleavages in Norwegian Politics' in Seymour M. Lipset and Rokkan (eds.), *Party Systems and Voter Alignments* (New York: Free Press, 1967).
10. Tarrow, *Between Center and Periphery*, and Tarrow, Katzenstein and Graziano (eds.), *Territorial Politics in Industrial Nations*.
11. See, for example, Milton J. Esman, *Ethnic Conflict in the Western World* (Ithaca: Cornell University Press, 1977); L. J. Sharpe, 'Devolution and Celtic Nationalism in the U.K.', *West European Politics*, Vol. 8 (1985), No. 3; Edward A. Tiryakin and Ronald Rogowski (eds.), *The New Nationalisms of the Developed West* (Boston: Allen & Unwin, 1985).
12. P. Anderson, *Passages From Antiquity to Feudalism* (London: New Left Books 1974); Anderson, *Lineages of the Absolutist State* (London: New Left Books, 1974); W. McNeill, *The Shape of European History* (New York: Oxford University Press, 1974); I. Wallerstein, *The Modern World System* (London: Academic Press, 1974).

13. Gottman, 'The Evolution ...', p. 31.
14. Max Weber, *The Theory of Social and Economic Organization* (London: Hodge, 1947), p. 143.
15. R. A. Dahl and E. R. Tufte, *Size and Democracy* (Stanford: Stanford University Press, 1974).
16. Ernest S. Griffith, 'Cultural Prerequisites to a Successfully Functioning Democracy', *American Political Science Review*, Vol. 50 (1958), No. 2; L. J. Sharpe, 'The Failure of Local Government Modernization in Britain: a Critique of Functionalism', *Canadian Public Administration*, Vol. 2 (1981), No. 1; Arthur B. Gunlicks (ed.), *Local Government Reform and Reorganization* (Port Washington, NY: Kennikat Press, 1981); Michael Hecter, *Internal Colonialism* (Berkeley: University of California Press, 1977).
17. Arend Lijphart, 'Consociational Democracy', *World Politics*, Vol. 21 (1969), No. 2, pp. 207–25; Val R. Lorwin, 'Segmented Pluralism: Ideological Cleavages and Political Cohesion in the Smaller European Democracies', *Comparative Politics*, Vol. 3 (1971), No. 2, pp. 141–75.
18. Katzenstein, *Corporatism and Change: Austria, Switzerland and the Politics of Industry* (Ithaca: Cornell University Press, 1984); Katzenstein, *Small States in World Markets: Industrial Policy in Europe* (Ithaca: Cornell University Press, 1985).
19. C. D. Foster, R. Jackman and M. Perlman, *Local Government Finance in a Unitary State* (London: Allen & Unwin, 1980), pp. 127–8.
20. Preston King, *Federalism and Federation* (London: Croom Helm, 1980), supra.
21. Sharpe, ' "Reforming" the Grass Roots: An Alternative Analysis', in David Butler and A. H. Halsey (eds.), *Policy and Politics* (London: Macmillan, 1978).
22. See especially, Rokkan, 'Dimensions of State Formation and Nation-building: a Possible Paradigm for Researching on Variations Within Europe' in C. Tilly (ed.), *The Formation of National States in Western Europe* (Princeton: Princeton University Press, 1975).
23. Rokkan and Urwin, Chap. 2.
24. Douglas C. North, *Structure and Change in Economic History* (New York: Norton, 1981), p. 8.
25. North, Chap. 2.
26. Ernest Gellner, *Nations and Nationalism* (Oxford: Blackwell, 1983), p. 99.
27. See William R. Beer, *The Unexpected Rebellion: Ethnic Activism in Contemporary France* (New York: New York University Press, 1980), p. 42.
28. J. S. Mill, *Considerations on Representative Government*, ed. R. McCallum (Oxford: Blackwell [1861], 1946), p. 294.
29. Sharpe, 'Decentralist Trends in Western Democracies: A First Appraisal', in Sharpe (ed.), *Decentralist Trends in Western Democracies* (London: Sage 198), pp. 9–79.
30. Beer, *The Unexpected Rebellion*.
31. See, for example, Robert Blauner, 'Internal Colonialism and the Ghetto Revolt', *Social Problems*, Vol. 16 (1969); P. G. Casanova, 'Internal Colonialism and National Development', *Studies in Comparative International Development*, Vol. 1 (1965), No. 4; Gellner, *Thought and Change* (London: Weidenfeld & Nicolson, 1964); Michael Hecter, *Internal Colonialism* (Berkeley: University of California Press, 1977); T. Nairn, *The Break-Up of Britain* (London: New Left Books, 1977) and Rudolfo Staverhagen, 'Classes, Colonialism and Acculturation', *Studies in Comparative International Development*, Vol. 1 (1965), No. 6.
32. For a reappraisal of the internal colonialism thesis, sometimes by those who originally argued for it, see Edward A. Tiryakin and Ronald Rogowski (eds.), *The New Nationalisms of the Developed West* (Boston: Allen & Unwin, 1985).
33. L. J. Sharpe, 'Devolution and Celtic Nationalism ...'
34. Tarrow, *Between Center*, p. 25.
35. Sharpe, 'The Failure', pp. 107–8.
36. Ibid., pp. 105–7.
37. Tilly (ed.), *The Formation of National States ...*
38. Bud B. Khleif, 'Issues of Theory and Methodology in the Study of Ethnolingual Movements', in Tiryakian and Rogowski, p. 90; Arend Lijphart, *Democracy in Plural Societies* (New Haven: Yale University Press, 1977).
39. Rokkan and Urwin, p. 144.
40. Meic Stephens, *Linguistic Minorities in Western Europe* (Llandysul: Gomerz, 1978), p. 567.
41. Stephens, p. 363.

42. See, for example, A. Vandenbosch and S. J. Eldersveld, *Government of the Netherlands* (Lexington: University of Kentucky, 1947); Lijphart; Gordon L. Weil, *The Benelux Nations* (New York: Holt, Reinhart, 1970); Christopher Bagley, *The Dutch Plural Society* (London: Oxford University Press, 1973); S. J. Eldersveld, J. Kooiman and T. van der Tak, *Elite Images of Dutch Politics* (Ann Arbor: University of Michigan Press, 1981).
43. H. Daalder, 'On Building Consociational Nations: the Cases of the Netherlands and Switzerland', *International Social Science Journal*, Vol. 23 (1971), No. 3, p. 363.
44. Elmer Plischke, *Micro States in World Affairs* (Washington: American Enterprise Institute, 1977).
45. Foster *et al.*, pp. 127–8.
46. Frank Gould and Firoozeh Zarkesh, 'Local Government Expenditures and Revenues in Western Democracies: 1960–1982', *Local Government Studies*, Vol. 12 (1986), No. 1, p. 36.
47. Ibid., p. 34.

ABSTRACTS

Territorial Politics in the United Kingdom: The Politics of Change, Conflict and Contradiction
R. A. W. Rhodes

This contribution attempts to demonstrate that differentiation, disaggregation and interdependence – characteristics subsumed under the label 'the differentiated policy' – are of equivalent importance to parliamentary sovereignty, cabinet government and prime ministerial power – characteristics conventionally attributed to 'the unitary state' – for the analysis of British government in general and territorial politics in particular. It does so by providing a framework for analysing territorial politics; describing the development of territorial politics between 1945 and 1985; and providing an explanation of the major changes in this period. It is concluded that post-war trends cannot be seen as the erosion of 'local autonomy' but are better described as the growth of interdependence between levels of government; the proliferation of ambiguous and confused relationships; and, as a result, the co-existence of fragmentation in the centre (based on functional policy networks) with the centralisation of each policy network.

France: The Construction and Reconstruction of the Centre, 1945–86
Yves Mény

Central-local relations in France have always been involved and conflictual. Linguistic minorities and under-developed regions have exercised strong and conflicting pressures to obtain both more resources and greater cultural autonomy from central government. The 36,000 small communes and the big towns have also pursued differing objectives. Yet, osmotic relations lie between local and national élites, and the presence of a powerful central state administration at local level has generally ensured policies of compromise, collaboration and pragmatism. The Socialist government reforms of 1981–86 are a clear illustration of this: far from being radical and socialist, they reflect the consensus within the political élites and the constant search for a mythical reconciliation between local diversity and republican uniformity.

The Federal Republic of Germany: From Co-operative Federalism to Joint Policy-making
Joachim Jens Hesse

In recent times, West German federalism has experienced an increasing interdependence of its different territorial units. This development from 'co-operative federalism' to 'joint policy-making' is characterised by the growing importance of shared functions and resources and by complicated processes of joint problem-solving. Despite a number of critical appraisals, however, the intergovernmental system of the West German state has proved to be surprisingly stable and adaptable to changed conditions, thus avoiding structural reforms which have characterised a number of other West European countries. This analysis tries to identify the special characteristics of the West German intergovernmental system, giving rise to this kind of flexibility. To do so, the classical analysis of federal-state relations is widened, by including the municipal levels, by overcoming the traditional static analysis of intergovernmental policy-making and by commenting on recent intergovernmental developments in the Federal Republic of Germany.

Italy: Territorial Politics in the Post-war Years: The Case of Regional Reform
Robert Leonardi, Raffaella Y. Nanetti and Robert D. Putnam

Territorial politics have become an important component in the Italian political system in the post-war period. The issue received significant attention during the debate on the formulation of the Republic Constitution between 1946 and 1947, but it was only partially implemented with the recognition of regional autonomy for only five out of Italy's 20 regions. However, the problem reappeared as an element in the formulation of reform measures drawn-up at the end of the 1960s in response to the popular demand for democratic reforms and the administrative system's need for rationalisation − i.e. the Italian welfare state system had become too top-heavy and inefficient due to the country's piecemeal approach to social provision. The key to this reform of decision-making as well as policy implementation was the institutionalisation of territorial politics through the introduction of regional government in the country's remaining 15 regions. Due to a series of political factors, regional government in Italy evolved from the weak form initially introduced by the central authorities in 1970 to a much more substantial centre of decision-making and policy implementation at the sub-national level in the 1975−77 period. This contribution pays particular attention to the events during the 1970s and tries to analyse the reason why the institutionalisation of territorial politics has had such a profound impact on the structure of Italian politics.

The Netherlands: A Decentralised Unitary State in a Welfare Society
Theo A.J. Toonen

This analysis provides the reader with a general overview of intergovernmental relations (IGR) in the so-called decentralised unitary state of the Netherlands. The first part contains a summary description of the main actors, the formal legal framework and some long-term developments which are of importance in understanding contemporary Dutch IGR. The second part deals with the analysis of the structure and politics of Dutch IGR.

Taking the presumed hierarchical structure of a unitary state as a point of departure, the topic has mainly been studied from a perspective which stresses centralisation, central control and local resistance as the main features. By way of contrast this analysis adopts an interorganisational or multi-actor perspective. If one looks at the operational level of the system of IGR it becomes clear that not only hierarchy and centralisation but also specialisation, differentiation and interdependency are the structural features which underlie the consensual nature of the politics of intergovernmental relations in Dutch welfare society.

Ireland: The Interplay of Territory and Function
T.J. Barrington

Modern Ireland is largely a creation of the nineteenth century in its institutions and its external politics with the United Kingdom, the United States and the Roman Catholic Church, with relations with the first two now being moderated by the new and growing impact of the European Community. Internal politics also largely derive from the nineteenth century, especially the major issue of Northern Ireland. Since independence in 1922 Irish government has practised extreme democratic centralism: a highly centralised and subordinate civil service moderated by functional, but still centralised, delegation to single-purpose agencies, which have proliferated; but with total subordination and stagnation of local government. The neglect of institutional development has led to a condition of some administrative disorder. Thriving on this has been a remarkable degree of political clientelism, or 'brokerage'. Many of the current troubles of Irish government can be attributed to a set of public institutions that fail to adapt themselves to a society that, generally, is decisively on the move: Irish centripetalism is approaching its nemesis.

The West European State: The Territorial Dimension
L.J. Sharpe

Despite a recent revival of interest in the state, territoriality remains something of a Cinderella of political science. Too often it is ignored and when not ignored may be disguised by 'horizontalisation' as a class substitute. Thus, at least three aspects of the modern democratic state are usually under-emphasised: the unique character of every polity derived from its unique location; the capacity of territorial allegiance to over-ride class party or occupational allegiance; finally, the impact of physical scale on the political process. Further aspects of territoriality in Western Europe are explored including centre–periphery relations, state fragmentation and sub-national structural modernisation.